PERFORMANCE METRICS FOR THE GLOBAL NUCLEAR DETECTION ARCHITECTURE

Abbreviated Version

Committee on Evaluating the Performance Measures and Metrics Development for the Global Nuclear Detection Architecture

Nuclear and Radiation Studies Board
Division on Earth and Life Studies

NATIONAL RESEARCH COUNCIL
OF THE NATIONAL ACADEMIES

THE NATIONAL ACADEMIES PRESS
Washington, D.C.
www.nap.edu

THE NATIONAL ACADEMIES PRESS 500 Fifth Street, NW Washington, DC 20001

NOTICE: The project that is the subject of this report was approved by the Governing Board of the National Research Council, whose members are drawn from the councils of the National Academy of Sciences, the National Academy of Engineering, and the Institute of Medicine. The members of the committee responsible for the report were chosen for their special competences and with regard for appropriate balance.

This study was supported by Contract No. HSHQDC-11-D-00009 between the National Academy of Sciences and the Department of Homeland Security. Any opinions, findings, conclusions, or recommendations expressed in this publication are those of the author(s) and do not necessarily reflect the views of the organizations or agencies that provided support for the project.

International Standard Book Number-13: 978-0-309-29014-2
International Standard Book Number-10: 0-309-29014-7

Additional copies of this report are available for sale from the National Academies Press, 500 Fifth Street, NW, Keck 360,Washington, DC 20001; (800) 624-6242 or (202) 334-3313; http://www.nap.edu.

Printed in the United States of America

THE NATIONAL ACADEMIES

Advisers to the Nation on Science, Engineering, and Medicine

The **National Academy of Sciences** is a private, nonprofit, self-perpetuating society of distinguished scholars engaged in scientific and engineering research, dedicated to the furtherance of science and technology and to their use for the general welfare. Upon the authority of the charter granted to it by the Congress in 1863, the Academy has a mandate that requires it to advise the federal government on scientific and technical matters. Dr. Ralph J. Cicerone is president of the National Academy of Sciences.

The **National Academy of Engineering** was established in 1964, under the charter of the National Academy of Sciences, as a parallel organization of outstanding engineers. It is autonomous in its administration and in the selection of its members, sharing with the National Academy of Sciences the responsibility for advising the federal government. The National Academy of Engineering also sponsors engineering programs aimed at meeting national needs, encourages education and research, and recognizes the superior achievements of engineers. Dr. C. D. Mote, Jr., is president of the National Academy of Engineering.

The **Institute of Medicine** was established in 1970 by the National Academy of Sciences to secure the services of eminent members of appropriate professions in the examination of policy matters pertaining to the health of the public. The Institute acts under the responsibility given to the National Academy of Sciences by its congressional charter to be an adviser to the federal government and, upon its own initiative, to identify issues of medical care, research, and education. Dr. Harvey V. Fineberg is president of the Institute of Medicine.

The **National Research Council** was organized by the National Academy of Sciences in 1916 to associate the broad community of science and technology with the Academy's purposes of furthering knowledge and advising the federal government. Functioning in accordance with general policies determined by the Academy, the Council has become the principal operating agency of both the National Academy of Sciences and the National Academy of Engineering in providing services to the government, the public, and the scientific and engineering communities. The Council is administered jointly by both Academies and the Institute of Medicine. Dr. Ralph J. Cicerone and Dr. C. D. Mote, Jr., are chair and vice chair, respectively, of the National Research Council.

www.national-academies.org

COMMITTEE ON EVALUATING THE PERFORMANCE MEASURES AND METRICS DEVELOPMENT FOR THE GLOBAL NUCLEAR DETECTION ARCHITECTURE

ARDEN BEMENT (*Chair*), Purdue University (retired), West Lafayette, Indiana
KELLEY COYNER, Northern Virginia Transportation Commission, Arlington, Virginia
MARTHA CRENSHAW, Stanford University, Stanford, California
JAMES S. DYER, University of Texas, Austin
ROGER L. HAGENGRUBER, University of New Mexico, Albuquerque
JOHN H. HOLMES, Port of Los Angeles, California
EDWARD H. KAPLAN, Yale School of Management, New Haven, Connecticut
JOHN MATTINGLY, North Carolina State University, Raleigh
GREGORY S. PARNELL, University of Arkansas, Fayetteville
DONALD PROSNITZ, Independent Consultant, Livermore, California
THOMAS C. SCHELLING, University of Maryland, College Park
DETLOF von WINTERFELDT, University of Southern California, Los Angeles

Staff

JENNIFER HEIMBERG, Study Director
ERIN WINGO, Senior Program Assistant
SHAUNTEÉ WHETSTONE, Senior Program Assistant (through July 8, 2013)

Preface

The Domestic Nuclear Detection Office (DNDO) within the Department of Homeland Security (DHS) is responsible for developing and coordinating a cross-agency strategy, the Global Nuclear Detection Architecture (GNDA), to detect, analyze, and report on nuclear and radiological materials that are out of regulatory control. The GNDA is a global activity that involves programs, people, and technical systems in the United States and many other countries. It was mandated by presidential directive (in 2005) and public law (in 2006).

DNDO and its federal partners issued the Global Nuclear Detection Architecture Strategic Plan in 2010 which describes the high-level goals and federal agency responsibilities for implementing the GNDA. The U.S. government intends to undertake an annual review of the GNDA strategic plan to assess its effectiveness and identify new requirements arising from changes in technology and/or the threat environment.

DNDO has asked for advice from the National Research Council on developing quantitative approaches for assessing the effectiveness of the GNDA. This advice will be used to improve the GNDA strategic plan during future annual review cycles.

The committee approached this study by first gaining an understanding of what is meant by the "global nuclear detection architecture." We reviewed its documentation (strategic plan, annual reviews, and the domestic implementation plan). We interviewed staff from DNDO and its many federal partners. We visited the Ports of Los Angeles (LA) and Long Beach and the LA Joint Regional Intelligence Center. The committee also invited other government agencies to tell us about their measures of effectiveness and performance metrics (see Appendix A for full listing of the briefings received by the committee).

The GNDA is a worldwide network of detection and reporting capabilities controlled by many different entities and funding lines. It is meant to protect against a wide range of adaptive and committed adversaries. Developing metrics to measure the effectiveness of such a complex system of systems is a difficult problem.

DNDO and its partner agencies have developed documentation and an initial accounting of the many existing federal programs and activities that support nuclear detection and reporting objectives. The committee has developed metrics and an analysis framework that may help guide DNDO and its GNDA partners from this initial accounting to developing a capability to measure the effectiveness of the overall GNDA.

However, it became clear during the course of the study that the lack of a lead architect and a centralized GNDA budget (see Observation 1) make it difficult for the GNDA to function as a system rather than a collection of programs. The decision to address this concern (e.g., to assign clear leadership through organizational change) rests with the U.S. government.

I would like to extend special thanks to Captain John Holmes for organizing and the Ports of LA and Long Beach for hosting the committee during one of our information-gathering sessions. Our visit to the ports and the surrounding facilities and discussions with numerous stakeholders allowed the committee to see a unique example of federal, state, and local agencies truly working together toward a common goal of protecting the nation against threats.

Arden Bement, *Chair*

Acknowledgments

The committee wishes to acknowledge and thank a number of individuals and organizations for their valuable contributions to this study:

The Department of Homeland Security (DHS) Domestic Nuclear Detection Office (DNDO) for its sponsorship of this study, and especially DNDO staff members Brendan Plapp, John Zabko, Kimberly Koeppel, Greg Haugan, and Richard Passow.

Those who gave presentations at the committee's information-gathering meetings provided insight and information that made this report possible (presentations are listed in Appendix A):

- Major General Julie A. Bentz, National Security Council
- Brendan Plapp, DNDO Architecture and Plans Directorate (currently with the National Defense University)
- John Zabko, DNDO Architecture and Plans Directorate
- Kevin Hart, DNDO Architecture and Plans Directorate
- Teri N. Leffer, Department of Energy, National Nuclear Security Administration, Second Line of Defense (NA-256)
- Steven Streetman, Data Architecture Solutions, Inc.
- Henry Willis, RAND Corporation
- David Kulp, Department of Defense
- Ernest Muenchau, DNDO Operation Support Directorate (OSD) (deceased)
- Colonel Robert Kolterman, Defense Threat Reduction Agency (DTRA)

- Brian Savage, DNDO OSD, Joint Analysis Center
- James Smith, Los Alamos National Laboratory
- Mark Oliphant, DNDO Red Team and Net Assessments Directorate
- J. J. Fisher, DNDO OSD
- Bernie Bogdan, Federal Bureau of Investigation (FBI)
- J. C. Wyss, State Department
- David Travers, U.S. Environmental Protection Agency
- Detective Jeff Shanaphy, Los Angeles Port Police
- Sgt. Peter Jackson, Los Angeles Sheriff's Department

The Ports of Los Angeles and Long Beach graciously hosted one of our information gathering sessions and opened their doors to the committee for touring purposes, as did the National Marine Exchange in Los Angeles and the Joint Regional Intelligence Center (JRIC). The staff and committee are very grateful for the hospitality shown at these locations. The committee thanks the following agencies for providing briefings at that meeting:

- FBI
- Long Beach Police Department
- Los Angeles County Sheriff's Department
- Los Angeles Fire Department
- Los Angeles Police Department
- U.S. Coast Guard
- U.S. Customs and Border Protection
- Los Angeles Port Police

The committee extends special thanks to the staff of the National Research Council for supporting this study. Staff members who have contributed to this effort are Jenny Heimberg (study director), Kevin Crowley (director of the Nuclear and Radiation Studies Board), Erin Wingo (senior program assistant), Shauntée Whetstone (senior program assistant), and Toni Greenleaf (financial and program associate).

This report has been reviewed in draft form by individuals chosen for their diverse perspectives and technical expertise, in accordance with procedures approved by the Report Review Committee of the National Research Council. The purpose of this independent review is to provide candid and critical comments that will assist the National Research Council in making its published report as sound as possible and will ensure that this report meets institutional standards for objectivity, evidence, and responsiveness to the study charge. The review comments and draft manuscript remain confidential to protect the integrity of the deliberative process. We thank the following individuals for their review of this report:

- Margaret Chu, M.S. Chu and Associates, LLC
- William Hagan, independent consultant
- Milton Levenson, independent consultant
- David Morton, University of Texas, Austin
- C. Paul Robinson, President Emeritus Sandia National Laboratories
- Tim Runyon, Illinois Emergency Management Agency (retired)
- Henry Willis, RAND Corporation

Although the reviewers listed above provided many constructive comments and suggestions, they were not asked to endorse the conclusions or recommendations, nor did they see the final draft of the report before its release. The review of this report was overseen by Chris G. Whipple, ENVIRON, and John F. Ahearne, Sigma Xi. Appointed by the National Research Council, they were responsible for making certain that an independent examination of this report was carried out in accordance with institutional procedures and that all review comments were considered carefully. Responsibility for the final content of this report rests entirely with the authoring committee and the institution.

Contents

Summary

The Global Nuclear Detection Architecture (GNDA) is described as "a worldwide network of sensors, telecommunications, and personnel, with the supporting information exchanges, programs, and protocols that serve to detect, analyze, and report on nuclear and radiological materials that are out of regulatory control."[1] The Domestic Nuclear Detection Office (DNDO), an office within the Department of Homeland Security (DHS), coordinates the development of the GNDA with its federal partners. DNDO has asked the National Research Council (NRC) for advice on how to develop performance measures and quantitative metrics that can be used to evaluate the overall effectiveness and report on progress toward meeting the goals of the GNDA. The statement of task for this study can be found in Box S-1 (also Appendix B). The GNDA is a complex system of systems meant to deter and detect attempts to unlawfully transport radiological or nuclear (RN) material.[2] It was established to enhance the U.S. government RN detection activities in response to the perceived increase of the risk of nuclear terrorism following the 9/11 attacks.[3] Multiple federal, international, state, local, tribal, and industrial entities participate in activities that

[1] See http://www.dhs.gov/architecture-directorate. Accessed August 1, 2013."Out of regulatory control" describes materials that are being imported, possessed, stored, transported, developed, or used without authorization by the appropriate regulatory authority, either inadvertently or deliberately (DHS, 2011b, Vol. I, p. 4).

[2] Radiological material is used in a radiological dispersion device (RDD); nuclear material is used in an improvised nuclear device (IND) or nuclear weapon. These are considered two distinct threats (e.g., http://www.dhs.gov/radiological-attack-what-it, http://www.dhs.gov/nuclear-attack-what-it). Accessed August 1, 2013.

[3] Security and Accountability for Every Port Act of 2006 (P.L. 109-347).

BOX S-1
Statement of Task

An ad hoc committee will conduct a study and prepare a report to the Domestic Nuclear Detection Office (DNDO) on quantitative approaches for evaluating the effectiveness of the Global Nuclear Detection Architecture (GNDA), specifically in the context of the following two tasks:

Task 1: Assess the feasibility of using performance measures and quantitative metrics for GNDA.

The committee should assess the feasibility of using performance measures and quantitative metrics for evaluating progress in meeting these performance goals. This assessment should consider the following factors:

- Definition of performance measures for each of the performance goals in the Strategic Plan.
- Definition of quantifiable performance metrics for each performance measure including, as appropriate, efficiency, output, and outcome-oriented performance measures.
- Identification of data to be used to quantify these performance metrics.
- Identification of methodologies to be used to collect and analyze these data.
- Specification of performance target values for assessing the effectiveness of each performance measure.

If the use of performance measures and quantitative metrics is determined to be feasible, the committee should, to the extent practical, recommend specific performance measures, metrics, and the other supporting information described in the list above for consideration by the DNDO.

If the use of performance measures and metrics is determined to be infeasible, the committee should recommend alternative evaluation approaches.

Task 2: Recommend approaches for evaluating the overall effectiveness of the GNDA.

The committee should specifically recommend

- Approaches for developing an overall analysis framework to assess the effectiveness of the GNDA in terms of its ability to detect, deny, confuse, and/or deter adversaries.
- Approaches for exercising this analysis framework using combinations of modeling/simulation, red teaming, and/or related methods to assess the cost-effectiveness of and tradeoffs in GNDA components.

In executing these tasks the committee should examine efforts by other organizations to develop risk-informed metrics and analysis approaches for complex technological systems.

can contribute to increasing the effectiveness of the global nuclear detection infrastructure.

The challenge presented to the GNDA federal partners, who are responsible through the Government Performance and Results Act (GPRA, P.L. 103-62) to report on the performance of the GNDA, is to develop meaningful metrics[4] and gather the appropriate data to gauge its overall progress every year. This challenge has also been considered carefully by this committee (see Task 1 in Box S-1). The committee is directed to assess the feasibility of developing measures and metrics against existing performance goals of the strategic plan that can be used to measure the effectiveness of the GNDA. If infeasible, the committee is to recommend alternative approaches to evaluating GNDA effectiveness.

There are significant challenges to developing metrics to gauge the overall effectiveness of the GNDA. The GNDA was created to address the threat of a high-consequence event that has never occurred. It must protect against a wide variety of adaptive and committed adversaries and threat materials. However, these types of challenges exist within other U.S. government agencies and are not unprecedented. The committee provides several examples and concludes that it is fundamentally *possible* to develop outcome-based metrics to gauge the effectiveness of the GNDA. However, the committee finds that it is not feasible to develop outcome-based metrics against the existing performance goals within the existing GNDA strategic plan. There are two reasons for this: the higher-level goals and objectives within the strategic plan are focused on process and activities (they are not primarily outcome-based) and many objectives are focused on individual GNDA layers or resources (not the full architecture); and the higher-level goals are disconnected from the objectives and lower-level performance goals. Furthermore, a new analysis framework is needed to evaluate the metrics and to prioritize the GNDA's goals and objectives. This report presents a notional strategic plan (with notional outcome-based metrics) and new analysis framework.

In addressing the statement of task, the committee identified several issues of concern that could limit the GNDA federal partners from implementing the findings and recommendations provided within this report. These concerns have been identified as observations.

A summary of observations, findings, and recommendations is provided below.

---

[4] Here and within the report, the committee uses the simplified "metric" to refer to "performance measures and quantitative metrics." While acknowledging that "measures and metrics" exist separately within the scholarly literature and measurement theory, for the committee's purposes in addressing the task statement, the simplified term "metric" is used.

OBSERVATIONS, FINDINGS, AND RECOMMENDATIONS

The committee recommends an approach for developing an updated strategic plan, outcome-based metrics, and an analysis framework for evaluating the effectiveness of the overall GNDA. The GNDA, however, currently does not function as an integrated system but as a collection of programs. If it is decided that the GNDA should function as a system, then it will need to have clearly-defined lead authority and a centralized budget so that reallocations can be made across programs and agencies. Nonetheless, the committee's recommendations for improved metrics and an analysis framework within the existing organizational structure could provide information to reallocate funds within agencies and to measure the overall cost of the GNDA.

OBSERVATION 1: There is no clear lead architect or single entity to make final decisions about or to be held accountable for the design and operation of the GNDA. Furthermore, there is no centrally controlled GNDA budget; GNDA-related detection and reporting activities are intertwined with diverse mission activities across GNDA federal agencies and do not have specific lines of funding. Thus, there is no single congressional appropriation for the GNDA nor is there a single entity with budgetary control over GNDA activities across multiple agencies.

The GNDA operates via a loosely confederated collection of federal, state/local and tribal programs and activities under what may be considered a "best-effort" budget. This is important to note, because it may not be possible to effectively utilize the results from an analysis framework and measures of effectiveness of the overall GNDA in a way that would change the contributions of participating agencies to the overall budget. This does not imply that developing improved metrics to guide resource decisions and establishing an analysis framework for the GNDA is without purpose. Establishing a capability to evaluate GNDA effectiveness can provide useful information to decision makers such as the gap between existing and optimal resource allocation and a measure of the cost of operating the GNDA. The issue of disconnected budgets' impact on coordination of the GNDA has been highlighted previously.[5] The committee provides an example of another federal program with similar challenges in Chapter 2.

OBSERVATION 2: The GNDA operates within a larger nuclear counterterrorism (NCT) mission. Its scope is limited to deterrence, detection, and

[5] This issue has been identified through Senate hearings (U.S. Congress, Senate, 2010 Hearing 111-1096) but no actions have been taken. (http://www.gpo.gov/fdsys/pkg/CHRG-111shrg58397/html/CHRG-111shrg58397.htm. Accessed on August 1, 2013).

reporting. When considering how to address and define the GNDA strategy and goals, focusing solely on the detection and reporting mission may limit wider U.S. government actions that span multiple components of the NCT mission space.

It is difficult to segregate actions and strategies focused on deterrence, detection, and reporting from other actions that support adjacent missions of federal agencies. The committee provides several examples of the impact of NCT federal mission boundaries on strategic planning and response options.

In Chapters 3 and 4, the committee provides one finding, one recommendation and notional examples of a strategic plan and outcome-based metrics in response to Task 1.

FINDING 1.1: **It is fundamentally possible to create outcome-based metrics for the GNDA; however, it is not currently feasible to develop outcome-based metrics against the existing strategic plan's goals, objectives, and performance goals because these components are primarily output- and process-based and are not linked directly to the GNDA's mission.**

Two conditions must be met to use metrics to evaluate the effectiveness of the GNDA:

1. **A new strategic plan with outcome-based goals and objectives must be created and**
2. **An analysis framework must be developed to enable assessment of outcome-based metrics.**

RECOMMENDATION 1.1:
When DNDO and the GNDA partner agencies next update the GNDA Strategic Plan, the committee recommends that they take the following steps:

1. **Generate a vision statement.**
   Without a clear, interagency-supported idea of the long-term goal of the GNDA, it is difficult to measure progress toward achieving it.
2. **Simplify the plan.**
   Limit the strategic plan's hierarchy to vision, mission, goals, and objectives; the goals and objectives should be outcome-based and they should clearly describe the desired results and how they are directly related to the mission and vision of the GNDA.
3. **Consider the broader nuclear counterterrorism problem before focusing on "detection."**

A strategic plan developed by solely focusing on deterrence, detection, and reporting mission may not fully consider the activities that take place at the mission interfaces. Therefore, a broader perspective is needed to initially determine strategic goals and objectives before they are limited to those within the scope of the GNDA.

4. **Determine the goals and objectives by focusing on the mission.**
   Do not limit the plan's goals and objectives by focusing on what can be easily measured or by what data are readily available. Some important objectives may not lend themselves to direct measurement but they should not be excluded from the plan for that reason.
5. **Use proxies when direct metrics are not available.**
   The metrics developed directly against outcome-based objectives will more readily be outcome-based and focused on measuring the full architecture. However, it is not always possible to develop metrics that meet these criteria. In those cases, proxies (i.e., indirect metrics that are frequently output- or process-based, such as the number of deployed detectors) can provide useful information as long as they can be directly linked to the objectives.

Furthermore, in the absence of a GNDA design document the committee suggests that the strategic plan clearly describes the GNDA's design goals and how its existence enhances the otherwise disparate detection activities of GNDA partner agencies.

In Chapter 5, the committee provides two findings, one recommendation and an example of a new analysis framework in response to Task 2.

FINDING 2.1:
A new GNDA Analysis Framework is needed to assess the GNDA effectiveness as shown in Figure 5-1. The critical components of the framework are the following:

1. **A GNDA Strategic Plan that contains outcome-oriented, broadly-scoped goals, objectives and metrics and is directly connected to the components listed below;**
2. **A GNDA Architectural Definition that provides the conceptual and physical descriptions of the GNDA, and that define needed input data for the models described below;**
3. **A suite of GNDA models that incorporate potential adversary objectives, accurately represent existing and potential architecture capabilities, and calculate the metrics described below;**
4. **Metrics that can gauge overall GNDA effectiveness and assess potential GNDA resource decisions to increase GNDA effectiveness; and**

5. A Validation and Verification (V&V) program that evaluates the data used in the GNDA architecture definition, models, and metrics. A robust V&V program enhances the credibility of the analysis framework.

FINDING 2.2:
Current DNDO modeling, testing, red teaming, analysis, and training capabilities provide a foundation for evaluating components of the GNDA, but these current capabilities are insufficient for validating and verifying the overall effectiveness of the GNDA. Evaluating the effectiveness of the overall GNDA requires an integrated and continuous model-based scenario testing, red teaming, analysis, peer review, and training supplemented with intelligence awareness.

RECOMMENDATION 2.1:
DNDO should develop a new GNDA Analysis Framework similar to the framework proposed by the committee. This framework defines an analytic process that clarifies the connections among strategic planning, architectural definition, models, metrics, and validation and verification efforts. Such an analysis framework can provide credible assessments of overall GNDA effectiveness.

1

Introduction

The Global Nuclear Detection Architecture (GNDA) is described as "a worldwide network of sensors, telecommunications, and personnel, with the supporting information exchanges, programs, and protocols that serve to detect, analyze, and report on nuclear and radiological materials that are out of regulatory control."[1] The Domestic Nuclear Detection Office (DNDO), an office within the Department of Homeland Security (DHS), coordinates the development of the GNDA with its federal partners. DNDO has asked the National Research Council (NRC) for advice on how to develop performance measures and quantitative metrics that can be used to evaluate the overall effectiveness and report on progress toward meeting the goals of the GNDA. The statement of task for this study can be found in Appendix B.

This chapter provides background, sponsor motivations, and the committee's approach to addressing the study charge.

1.1 BACKGROUND

U.S. government programs focused on the detection of nuclear and radiological materials have existed for many years but for the most part have been developed independently of each other. In response to the increased threat of nuclear terrorism, the GNDA was introduced by Presidential

[1] See http://www.dhs.gov/architecture-directorate. Accessed August 1, 2013.

Directive[2] in 2005 as an integrated and coordinated architecture of U.S. nuclear detection assets around the world. DNDO, simultaneously created by the same directive, was assigned to coordinate GNDA development and implement its domestic portion.

In 2006, the Security and Accountability for Every (SAFE) Port Act (P.L. 109-347), established the DNDO under DHS, created the enhanced GNDA, and assigned DNDO the responsibility of GNDA development as the coordinating agency (emphasis added in the text below):

> **[DNDO shall] develop**, with the approval of the Secretary of Homeland Security and in coordination with the Attorney General and Secretaries of State, Defense, and Energy, **an enhanced global nuclear detection architecture** with the following implementation—
>
> (A) [DNDO] will be responsible for the implementation of the domestic portion of the global architecture;
>
> (B) the Secretary of Defense will retain responsibility for implementation of Department of Defense requirements within and outside the United States; and
>
> (C) the Secretaries of State, Defense, and Energy will maintain their respective responsibilities for policy guidance and implementation of the portion of the global architecture outside the United States, which will be implemented consistent with applicable law and relevant international arrangements (NSPD-43, 2.d)

In 2007, Congress amended the SAFE Port Act (P.L. 110-53) requiring that a GNDA Joint Annual Interagency Review be provided to Congress, the President and the Office of Management and Budget:

> JOINT ANNUAL INTERAGENCY REVIEW OF GLOBAL NUCLEAR DETECTION ARCHITECTURE . . . .
>
> . . . jointly ensure interagency coordination on the development and implementation of the global nuclear detection architecture by ensuring that, not less frequently than once each year—
>
> (A) each relevant agency, office, or entity—
>
> (i) assesses its involvement, support, and participation in the development, revision, and implementation of the global nuclear detection architecture; and
>
> (ii) examines and evaluates components of the global nuclear detection architecture (including associated strategies and acquisition plans) relating to the operations of that agency, office, or entity, to

[2] Both DNDO and the GNDA were initially established on April 15, 2005 via National Security Presidential Directive 43 (NSPD-43) and Homeland Security Presidential Directive HPSD-14. (http://www.dhs.gov/architecture-directorate. Accessed August 1, 2013.)

> determine whether such components incorporate and address current threat assessments, scenarios, or intelligence analyses developed by the Director of National Intelligence or other agencies regarding threats relating to nuclear or radiological weapons of mass destruction; and
>
> (B) each agency, office, or entity deploying or operating any nuclear or radiological detection technology under the global nuclear detection architecture—
>
> (i) evaluates the deployment and operation of nuclear or radiological detection technologies under the global nuclear detection architecture by that agency, office, or entity;
>
> (ii) identifies performance deficiencies and operational or technical deficiencies in nuclear or radiological detection technologies deployed under the global nuclear detection architecture; and
>
> (iii) assesses the capacity of that agency, office, or entity to implement the responsibilities of that agency, office, or entity under the global nuclear detection architecture. (6 USC § 596a)

There are no official design documents that provide a systems-level description of the GNDA. However, there are several programmatic documents that define its mission and describe aspects of its design. DNDO and its federal partners have produced several key GNDA-related documents and presentations that describe aspects of the overall architecture to accomplish this mission (all of the following documents are restricted from public access):

- The GNDA Strategic Plan (GNDA, 2010);
- Three Joint Annual Interagency Reviews (GNDA, 2010, 2011, 2012);
- The Department of Homeland Security GNDA Implementation Plan (DHS, 2012), which addresses the domestic portion of the GNDA; and
- Presentation of the Draft International GNDA Implementation Plan (Wyss, 2012).

The Government Accountability Office (GAO) identified the need for a strategic plan to guide development of the GNDA (GAO, 2008). In recent testimony to Congress, GAO credits DNDO and its federal partners for developing both the strategic and implementation plans but notes that DNDO needs to prioritize the various objectives related to domestic activities:

> We reported, in July 2011, that the GNDA strategic plan addressed several of the aspects of our prior recommendations but did not (1) identify funding necessary to achieve plan objectives or (2) employ monitoring

> mechanisms to determine progress and identify needed improvements. In April 2012, DHS issued its GNDA implementation plan, which addresses the remaining aspects of our recommendations by identifying funding dedicated to plan objectives and employing monitoring mechanisms to assess progress in meeting those objectives. However, in both the GNDA strategic plan and the implementation plan, it remains difficult to identify priorities from among various components of the domestic part of the GNDA. (GAO, 2012, p. 3)

This study addresses the concerns raised by the GAO by considering how the development of new metrics and a new analysis framework could be used to optimize and prioritize GNDA resources.

1.2 MOTIVATION FOR THIS STUDY

DNDO and its GNDA-partner agencies intend to undertake a periodic review of the GNDA strategic plan to assess its effectiveness and identify new requirements arising from changes in technology and/or the threat environment. As noted previously, DNDO has asked for advice from the NRC on developing quantitative approaches for assessing the effectiveness of the GNDA. This advice may be used to improve the GNDA strategic plan and the reporting of progress toward meeting its goals during subsequent review cycles.

Currently, DNDO collects information for the joint annual interagency review from each GNDA partner agency which assesses its own involvement. Combining these and other data to provide an overall assessment of GNDA effectiveness, rather than a listing of individual programs, is one of the main challenges for this study.

There are many other challenges for evaluating the effectiveness of the GNDA. The GNDA was created to address the threat of a high-consequence event that has never occurred. It must protect against a wide variety of adaptive and committed adversaries and threat materials. As will be discussed later in the report, evaluating and comparing probabilities of different attack scenarios can be used to address this complex problem but this (and all other approaches) have the challenge of characterizing the full universe of potential attack pathways.

1.3 COMMITTEE'S APPROACH TO ADDRESS THE STUDY CHARGE

This study was carried out by a committee of 12 experts appointed by the NRC. The committee's collective expertise spans the issues relevant to the study task: cost-benefit analysis, decision analysis (especially multi attribute utility analysis), risk analysis, national security, nuclear materials characteristics and behavior, program evaluation and assessment, strategic planning, systems analysis, and technology development and deployment. In selecting the membership of this committee, the NRC sought to obtain a balance between members with relevant disciplinary expertise and subject-matter experts who have on-the-ground experience with testing and evaluating complex technological systems and multiorganization programs. Biographical sketches of the committee members are provided in Appendix C.

Through discussions with DNDO and its GNDA partner agencies, the committee determined that the focus of the study (see Appendix B) is the development and definition of appropriate metrics and an evaluation framework that can be used to assess and report on the overall effectiveness of the GNDA. This report is not an assessment of how effectively the current GNDA is performing. The committee was not asked to evaluate the DNDO or its partner agencies, to assess the existing organizational or budgetary structure, or to develop an implementation plan. Rather, the report provides notional metrics and an analysis framework that can be used to evaluate the effectiveness of the GNDA. Nonetheless, the committee needed to develop a detailed understanding of the current GNDA, its organization and funding mechanisms, the GNDA strategic plan, and the annual review process to make recommendations on how to measure and report on its overall effectiveness.

The committee held seven meetings over 12 months, including site visits to the Ports of Los Angeles and Long Beach. The committee received briefings at five of the meetings from DNDO and its partners in the GNDA, other government agencies with established metrics and measures for security-related missions, and researchers investigating complex security systems. A list of meetings and presentations is provided in Appendix A.

1.4 REPORT ROADMAP

The report is organized into five chapters:

- Chapter 1 provides the background, study charge, and structure for the report.
- Chapter 2 describes the GNDA in terms of its scope, participants, and structure. General observations are provided on challenges to the GNDA.
- Chapter 3 addresses Task 1. Key terms and definitions are provided as well as criteria for development of informative and useful metrics.
- Chapter 4 describes a notional strategic planning example that includes outcome-based metrics.

- Chapter 5 addresses Task 2 and introduces an analysis framework to evaluate the overall effectiveness of the GNDA.

An effort was made by the committee to develop chapters that could stand alone for the benefit of audiences who were not interested in reading the entire report. This results in some repetition of basic facts and concepts in chapters that will be noticed by those who read the report from beginning to end.

2

GNDA Background

This chapter provides information about the Global Nuclear Detection Architecture (GNDA). The first section describes the GNDA: specifically, what it is and who its partners are. The second section describes the challenges to evaluating its effectiveness.

2.1 GNDA DESCRIPTION

The committee spent a considerable amount of time understanding the design and structure of the GNDA. To the committee's knowledge there is no single document that provides a detailed description of the functions, requirements, and design of the GNDA. However, the GNDA Strategic Plan (GNDA, 2010) and Joint Annual Interagency Review (GNDA, 2011, 2012) do provide general descriptions of the design. In addition to these documents, the committee received briefings from DNDO and its federal, state, and local partners that helped to complete its understanding of the GNDA as it currently exists. These briefings are listed in Appendix A.

According to the Department of Homeland Security (DHS), the GNDA is "a worldwide network of sensors, telecommunications, and personnel, with the supporting information exchanges, programs, and protocols that serve to detect, analyze, and report on nuclear and radiological materials that are out of regulatory control."[1] The GNDA is a complex system of systems involving many U.S. and international organizations whose collective purpose is to reduce the risk of radiological or nuclear terrorist attacks

[1] See http://www.dhs.gov/architecture-directorate. Accessed August 1, 2013.

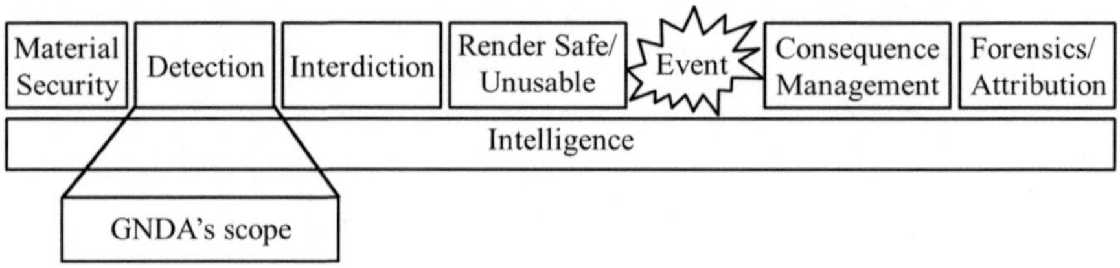

FIGURE 2-1 The nuclear counterterrorism (NCT) operational spectrum is described from the origin or location of radiological or nuclear material through detonation ("the boom") to forensics and attribution. Within this spectrum, the GNDA's mission scope occurs between material security (RN material under regulatory control) and interdiction (return to regulatory control).
SOURCE: Modified from National Nuclear Security Administration (NNSA; http://nnsa.energy.gov/aboutus/ourprograms/ctcp/nuclearthreatscience. Accessed August 1, 2013).

through detection and reporting capabilities. The GNDA defines detection to include both technical (detection equipment) and nontechnical (an information alert) means.[2] The GNDA is often described as having a defense-in-depth structure organized by groups of nuclear detection capabilities distributed across three geographical layers (a layer external to the United States; a transborder layer; and an interior layer) and a fourth crosscutting layer (such as intelligence, coordination, and communication functions).

The GNDA's detection capabilities are designed and deployed to detect radiological and nuclear material outside of regulatory control. They are not designed to detect an unauthorized nuclear detonation or test. Such capabilities are outside the scope of the GNDA and are the responsibility of other agencies and programs. They were not investigated in this study. The GNDA is one part of the U.S. nuclear counterterrorism (NCT) mission to reduce the risk of nuclear and radiological terrorist or covert host-state attacks. The NCT mission is frequently displayed as a spectrum of operational activities that occur either before or after a nuclear or radiological event (see Figure 2-1). Activities related to the NCT are referenced within this community as "left of the boom" and "right of the boom." The scope of the GNDA, to detect and report on occurrences of radiological and nuclear (RN) material discovered out of regulatory control, is left of the boom and within the "detection" portion of the spectrum. Interdiction of nuclear material (e.g., recovery of material) is not part of the GNDA's mission, nor is material security (e.g., physical security of nuclear materials and the facilities that produce them) but both activities interface directly with the GNDA. Furthermore, the scope of the GNDA does not include intel-

[2] See Appendix F for a glossary of terms.

ligence functions such as threat definition. These functions are performed by and shared through the Director of National Intelligence (DNI).[3]

Within this spectrum, the GNDA's scope can be considered as being analogous to "bell ringer" systems, such as the worldwide tsunami detection system, that serve to discriminate between false alarms and actual events and provide warnings of real threats to the appropriate partners in actionable time frames (NTHMP, 2013). However, this analogy excludes an important component of the GNDA's mission that makes it distinct from natural disasters such as tsunamis. The GNDA is preventive and includes two components: (1) deterring an adversary and, if that fails, (2) detecting and reporting of undeterred attempts.

DNDO and its federal partners within the GNDA seek ways to assess the effectiveness of the GNDA against the threat of intelligent, adaptive adversaries—including the effectiveness of deterrence. Deterrence and its characterization as they relate to the GNDA are discussed in Chapter 4.

2.1.1 GNDA Participants

DNDO, in its coordination role for the GNDA, works closely with several federal nuclear security partners, including

- Department of Energy (DOE) National Nuclear Security Administration (NNSA),
- DNI,
- Department of Defense (DOD),
- Department of State (DOS),
- Department of Justice, primarily the Federal Bureau of Investigation (FBI),
- Nuclear Regulatory Commission (USNRC), and
- other DHS agencies including the U.S. Coast Guard, Customs and Border Protection, and the Transportation Security Administration.

However, DNDO does not have the lead role for either designing or controlling the GNDA. In fact, no single agency or entity has a clearly defined lead.

Additional participants in the GNDA include international, federal (in addition to the GNDA partners listed above), state, local, tribal, and territorial entities.[4] DHS and DNDO do not have direct authority over these participants. Furthermore, these entities are not obligated to participate in GNDA activities. In many instances these participants might not be aware

[3] P.L. 110-53.

[4] See http://www.dhs.gov/architecture-directorate. Accessed August 1, 2013.

that their activities are considered part of the GNDA, and they may not understand the GNDA acronym, mission, or scope. These partners are not as familiar with the term "GNDA" as they are with "preventive radiation and nuclear detection" (PRND) programs and activities that span multiple parts of the federal NCT spectrum (see Figure 2-1).[5] International partners are also not obligated to participate in the GNDA. Efforts to work cohesively and within mission goals of the GNDA with other countries are coordinated by a group of federal agencies (as defined by the SAFE Port Act).

The GNDA does not have a central budget. Funding for GNDA-related activities is provided through a variety of federal agencies' appropriations bills or grants to state and local jurisdictions. Funding for nuclear detection capabilities (detectors, training, analysis, and alerting) is provided directly to GNDA participants usually as part of funding for a larger mission. Since there is no central GNDA budget, there is no central budgetary authority or oversight control.[6]

2.1.2 GNDA Structure

There are two main views of the GNDA structure. The first is the "geographical view" of the GNDA; it is described as a set of three main geographical layers (see Figure 2-2).

The description of each layer can be found on DHS's website.

- "The interior layer of the GNDA includes all areas within and up to, but not including, the U.S. border. The interior layer focuses on increasing nuclear detection capabilities across the maritime, air, and land pathways and addressing a wide array of potential threats."[7] Under the SAFE Port Act, DNDO is responsible for implementation of the domestic portion of the GNDA.
- "The transit and border layer (trans-border) is composed of transit to the United States from a foreign port of departure or non-port of departure, as well as passing through the U.S. border prior to entering the U.S. interior. This represents the last opportunity to detect radiological or nuclear materials prior to their arrival onto U.S.

[5] The article, "Preventing the Theft of Dangerous Radiological Materials," by Edward Baldini (2010) describes the Philadelphia Police Department's PRND activities with DNDO and other federal agencies without mentioning the GNDA. (http://www.policechiefmagazine.org/magazine/index.cfm?fuseaction=display_arch&article_id=2199&issue_id=92010. Accessed August 1, 2013.)

[6] This issue has also been identified through Senate hearings (U.S. Congress, Senate, 2010) but no actions have been taken. (http://www.gpo.gov/fdsys/pkg/CHRG-111shrg58397/html/CHRG-111shrg58397.htm. Accessed on August 1, 2013).

[7] Layered Nuclear Defense: Interior Layer, http://www.dhs.gov/layered-nuclear-defense-interior-layer. Accessed August 1, 2013.

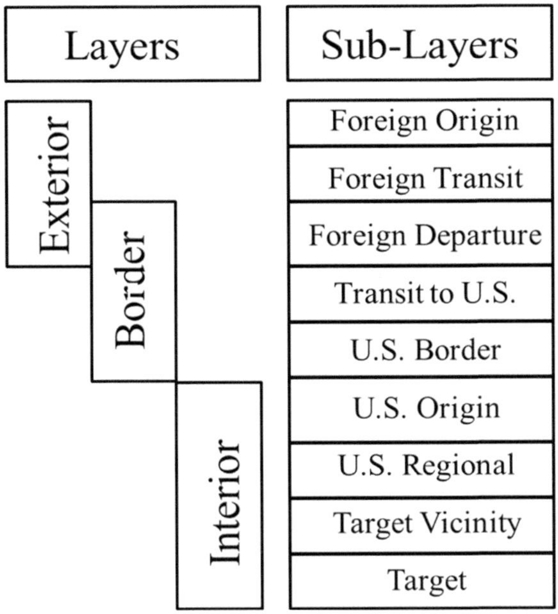

FIGURE 2-2 Geographical view of the GNDA. In this view, the GNDA is described as having three geographical layers (exterior, border, and interior). The crosscutting functions are not shown.
SOURCE: Modified from DHS (2007).

territory, and initiatives in this layer emphasize maritime domain awareness related to preventive radiological/nuclear detection."[8] Although maritime domain awareness is highlighted in this text, land and air transportation pathways are considered part of the GNDA.

- "The exterior layer comprises the foreign origin, foreign transit and foreign departure sub-layers. We improve radiological and nuclear material detection abroad through efforts that encourage foreign nations or regions to develop and enhance their nuclear detection architectures."[9] Under the SAFE Port Act, DOS, DOE, and DOD are responsible for implementation of the exterior portion of the GNDA consistent with international agreements and laws.
- "Cross-cutting efforts focus on programs and capabilities spanning

[8] Layered Nuclear Defense: Trans-border Layer, http://www.dhs.gov/layered-nuclear-defense-trans-border-layer. Accessed August 1, 2013.

[9] Layered Nuclear Defense: Exterior Layer, http://www.dhs.gov/layered-nuclear-defense-exterior-layer. Accessed August 1, 2013.

> multiple layers and pathways of the GNDA. Efforts undertaken in this layer provide the basis for time-phased deterrence and detection strategies. These elements streamline existing capabilities, improve overall coordination and ultimately seek to enhance radiological and nuclear detection at the federal, state, territorial, tribal and local levels."[10]

In the geographical, three-layered view of the GNDA, transportation pathways and detection capabilities are grouped into modalities (e.g., land, air, sea for pathways; passive radiation portals, or handheld sensors for detection capabilities) with combinations of modality pathways and capabilities considered against known aspects of the terrorist threat.[11]

The other view of the GNDA structure is an operational view (OV). A notional diagram of the GNDA OV is shown in Figure 2-3.[12] This view, when populated with the specific geographical locations of threats, capabilities, and targets, can provide an intuitive picture of current operational capabilities and redundancies across the GNDA. As was shown in the geographical view, gaps in operational coverage can also be identified and prioritized through threat analysis, and particular routes can be highlighted (e.g., pedestrians traveling from Mexico to the United States between the ports of entry in El Paso and Presidio, Texas). See Chapter 5 for a more detailed discussion on these two views of the GNDA and their corresponding models. The global aspect of the architecture is clear in both views, so it is important to note that threats originating domestically are also included in the GNDA.

2.2 DISCUSSION

This report is not an assessment of how effectively the current GNDA is performing. The committee was not asked to evaluate the DNDO, its partner agencies, or the existing GNDA organizational or budgetary structure. This report addresses the study charge by providing examples of notional metrics and an analysis framework that can be used to evaluate the effectiveness of the GNDA. However, based on its understanding of the GNDA, the committee identifies several challenges that could affect the ability of

---

[10] Layered Nuclear Defense: Cross-cutting Efforts, http://www.dhs.gov/layered-nuclear-defense-cross-cutting-efforts. Accessed August 1, 2013.

[11] Threat characteristics are determined by the intelligence community. Threat assessment is outside the scope of the GNDA.

[12] The committee notes that this notional image is not an OV by the military terms. A military OV is a document that describes each node (e.g., land point of entry, or potential target) and its interaction with other nodes.

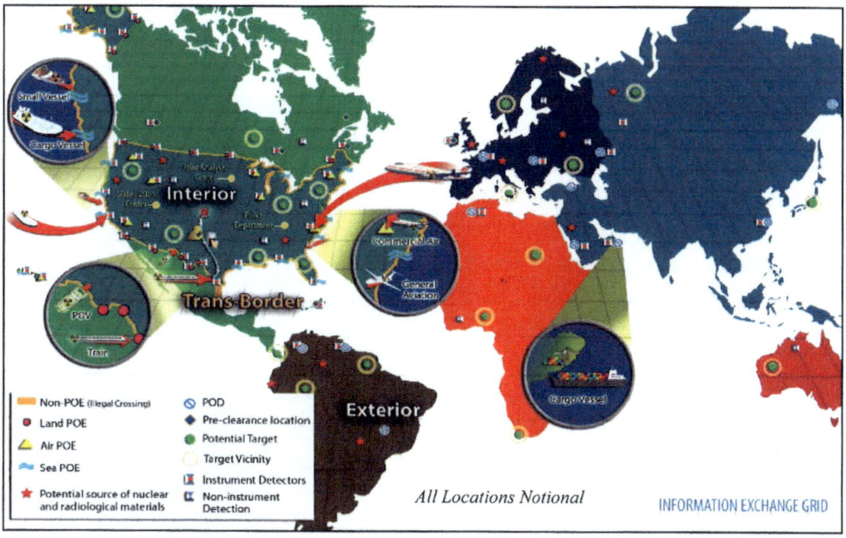

FIGURE 2-3 Operational View of the GNDA. In this view, existing threats and targets, transportation pathways, and current detection capabilities are mapped onto their actual geographic locations. This example is notional.
SOURCE: GNDA (2011).

DNDO and its GNDA partners to implement this report's findings and recommendations.

2.2.1 GNDA Governance

The public laws that created the GNDA do not assign it clear leadership. DNDO is designated as the coordinating entity and is frequently considered responsible for the GNDA (GAO, 2008; Shea, 2008). But there is no defined lead architect—whether an agency, entity, or person—to make decisions about or to be held accountable for design and implementation of the GNDA.

Furthermore, as mentioned earlier, there is no centralized GNDA budget. Funding is provided to multiple agencies for GNDA-related activities through multiple appropriations bills. Actual costs for activities can be difficult to estimate because detection and reporting of radiological and nuclear material out of regulatory control are part of larger missions executed by many partners. This introduces uncertainties and inconsistencies in the annual reported budget values. Without a clear understanding of the costs and the authority to make decisions, prioritization across the GNDA is very difficult. However, this situation is not unprecedented within the U.S. government.

Other organizations such as the National Earthquake Hazard Reduction Program (NEHRP) have addressed similar challenges. NEHRP is a set of four federal agencies[13] with separate budgets: "There is no single congressional appropriation for NEHRP, nor does the NEHRP Secretariat control individual agency budgets, personnel, or activities" (NEHRP, 2008, p. 12). Like the GNDA, NEHRP was established by law (Earthquake Hazards Reduction Act of 1977, as amended, 2004 [P.L. 95-124, 42 U.S.C. §§ 7701 et seq.][14]). Well-defined leadership was established by the NEHRP Reauthorization Act of 2004 (P.L. 108-360)[15] which established the NEHRP Interagency Coordinating Council (ICC). The ICC oversees NEHRP planning, management, and coordination and has the responsibility of developing the strategic plan. Members of the ICC include the White House Office of Science and Technology Policy, the Office of Management and Budget, and the directors of each of the four agencies that compose the NERHP; the ICC is chaired by the Director of NIST. The NEHRP agencies work closely to make decisions that mutually benefit the overall (and overlapping) mission when possible (NEHRP, 2008). The NEHRP strategic plan lists the agencies' coordinated vision, mission, goals, and objectives, but the implementation is the responsibility of each agency.

2.2.2 Critical Activities at Mission Boundaries

Different federal agencies are responsible for the different activities within the NCT mission spectrum (see Figure 2-1). Critical activities and decisions are made at the boundaries of these missions, which can lead to segmented agency activities and processes. The limited scope of "detection" was noted by J. C. Wyss (2012) in his presentation to the committee: "Nuclear detection is not a distinct event (p. 9)." The segmentation could affect the federal government's ability to fully consider strategies to combat threats, to fully integrate activities, and to coordinate exercises and lessons learned that cross mission boundaries.

The scope of the GNDA mission is detection of materials out of regulatory control (see Figure 2-1); the mission boundary to the left is "material security," and the mission boundary to the right is "interdiction." In the

---

[13] The four federal agencies are: Federal Emergency Management Agency (http://www.fema.gov/earthquake) of the Department of Homeland Security, National Institute of Standards and Technology (NIST, http://www.nist.gov/index.html) of the Department of Commerce (NIST is the lead NEHRP agency), National Science Foundation (http://www.nsf.gov/) and the United States Geological Survey (http://www.usgs.gov/) of the Department of the Interior.

[14] See http://www.nehrp.gov/about/PL108-360.htm.

[15] See http://www.nehrp.gov/about/PL108-360.htm.

sections below, the committee considers activities at each interface focusing on domestic examples to highlight U.S. federal agency involvement. Security of domestic sources of RN material is the responsibility of several federal agencies including USNRC, FBI, and NNSA. Interdiction within the United States is the responsibility of the FBI.

Material Security Boundary

The committee notes that significant progress has been made by the FBI and NNSA on providing training and exercises to secure materials at domestic facilities housing potential radiological dispersion device (RDD) threat material.[16] Box 2-1 has a detailed discussion on the differences between radiological and nuclear attacks. Tabletop exercises (e.g., the Silent Thunder tabletop series) include participation by the FBI, NNSA, state and local law enforcement, and industrial partners. The exercises are aimed at giving federal, state, and local officials, first responders, and law enforcement critical, hands-on experience in responding to a terrorist attack involving radiological materials (NNSA, 2012a). Communication and coordination on the concept of operations (CONOPS) developed within these multiple exercises can be specific to the state/local or industrial location. The results of these exercises which are focused on the mission of physical security of radiological sources are best shared across the federal mission boundaries (e.g., to include detection of radiological material out of regulatory control) so that they are seamless from the perspectives of state and local entities (see Box 2-1).

Interdiction Boundary

Critical activities occur and decisions are made at the interface of adjacent activities within the NCT mission. Federal responsibilities change hands at these interfaces, for example, the detection–interdiction interface. This could have an unintended consequence of limiting the U.S. government's choices in responding to a confirmed detection event. Clearly the operational decisions and subsequent actions that occur between confirmation of the detection of a threat material and its interdiction need to be made quickly (e.g., detection of a threat in a truck at a border crossing).

---

[16] "In the event that terrorists were able to obtain radiological materials and attempt to use them in an attack, NNSA has worked with federal, state, and local officials across the country through a series of tabletop exercises that strengthen first responders' and law enforcement officials' ability to detect, deter and prevent a terrorist WMD incident from occurring, as well as emphasize efforts to respond to, mitigate and recover from the effects of such an event. NNSA's 100th exercise of its kind was held in August" (NNSA, 2012b).

BOX 2-1
Radiological and Nuclear Attacks

Radiological and nuclear attacks are very different. DHS, in conjunction with the NRC, has defined both types of attack:

> A radiological attack is the spreading of radioactive material with the intent to do harm. Radioactive materials are used every day in laboratories, medical centers, food irradiation plants, and for industrial uses. If stolen or otherwise acquired, many of these materials could be used in a "radiological dispersal device" (RDD). (NAE/NRC, 2004)

> A nuclear bomb creates an explosion that is thousands to millions of times more powerful than any conventional explosive. . . . The resulting mushroom cloud from a nuclear detonation contains fine particles of radioactive dust and other debris that can blanket large areas (tens to hundreds of square miles) with "fallout.". . . The primary obstacle to a nuclear attack is limited access to weapon-grade nuclear materials. Highly enriched uranium, plutonium, and stockpiled weapons are carefully inventoried and guarded. (NAE/NRC, 2005)

A radiological or "dirty bomb" attack employs an RDD that uses means such as chemical explosives, for example, to widely disperse radiological materials. The radiological materials vary in source, isotope composition, and radioactivity level. The United States has many medical and industrial facilities that store and regularly use radiological materials. Because terrorists are more likely to seek sources within the United States to avoid long transportation routes, theft and misuse of radiological sources are serious threats. An RDD attack is listed as one of 15 disasters within National Planning Scenarios (DHS, 2006).

In contrast, a nuclear attack employs weapon-grade nuclear materials (highly-enriched uranium [HEU] and plutonium). The number of facilities storing weapon-grade materials is significantly less than those storing radiological sources. Within the United States, these materials are highly secured at a limited number of sites. Weapon-grade material is also stored at foreign facilities. Because weapon-grade materials are relatively scarce compared with radiological sources and because the impact of a nuclear attack is so large, it is thought that terrorists will attempt to obtain these materials wherever possible. In this case, the threat is not focused on domestic facilities but is considered global. An improvised nuclear device (IND) attack is identified as a scenario distinct from an RDD attack in the National Planning Scenarios list (DHS, 2006).

For the GNDA, these two attack modes represent separate overall architectures (Rosoff and von Winterfeldt, 2007). Preventing nuclear attacks puts an emphasis on securing foreign facilities and detecting nuclear materials en route to the United States. In contrast, preventing RDD attacks puts an emphasis on securing facilities in the United States and possibly establishing detection capabilities at major facilities (e.g., blood or food irradiation facilities). Although the physical security of sources is outside the scope of the GNDA, it has a direct impact on the evaluation of overall risk of a radiological attack.

There are two response options to a confirmed detection of threat material out of regulatory control:

1. The threat materials are returned promptly to control status upon confirmed detection and reporting, and
2. The detected threat materials are allowed to pass "seemingly undetected" to root out covert terrorist cells and networks, which may have the capabilities to transport and accumulate fissile and radiological materials and assemble, place, and detonate a nuclear device or RDD within the continental/contiguous United States.

The first case has been exercised repeatedly by the U.S. government. However, the second case demonstrates the challenge of making a decision about what to do with the detection information. Who would make a decision to not interdict? If it is not interdicted, which mission space does the activity now fall under? One could argue whether or not this scenario is realistic based on current capabilities and policies, but it provides an example in which the structure and responsibilities of the NCT mission space may have an unintended consequence of limiting U.S. government response options.

2.3 COMMITTEE'S OBSERVATIONS

The following observations are made to highlight potential challenges in implementing the committee's findings and recommendations which appear elsewhere in this report:

OBSERVATION 1: There is no clear lead architect or single entity to make final decisions about or to be held accountable for the design and operation of the GNDA. Furthermore, there is no centrally controlled GNDA budget; GNDA-related detection and reporting activities are intertwined with diverse mission activities across the GNDA federal agencies and do not have specific lines of funding. Thus, there is no single congressional appropriation for the GNDA nor is there a single entity with budgetary control over GNDA activities across multiple agencies.

The GNDA operates via a loosely confederated collection of federal, state/local and tribal programs and activities under what could be considered a "best-effort" budget. This is important to note, because it may not be possible to effectively utilize the results from an analysis framework and measures of effectiveness of the overall GNDA in a way that would change the contributions of participating agencies to the overall budget. This does not imply that developing improved metrics to guide resource decisions and establishing an analysis framework for the GNDA is without purpose.

Establishing a capability to evaluate the GNDA effectiveness can provide useful information to decision makers such as the gap between existing and optimal resource allocation and a measure of the cost of operating the GNDA. The issue of disconnected budgets' impact on coordination of the GNDA has been highlighted previously.[17]

OBSERVATION 2: The GNDA operates within a larger nuclear counter-terrorism (NCT) mission. Its scope is limited to deterrence, detection, and reporting. When considering how to address and define the GNDA strategy and goals, focusing solely on the detection and reporting mission may limit wider U.S. government actions that span multiple components of the NCT mission space.

It is difficult to segregate actions and strategies focused on deterrence, detection, and reporting from missions of federal agencies (Wyss, 2012). In the sections above the committee provides several examples of the impact of NCT federal mission boundaries on strategic planning and response options.

[17] This issue has been identified through Senate hearings (U.S. Congress, Senate, 2010 Hearing 111-1096) but no actions have been taken. (http://www.gpo.gov/fdsys/pkg/CHRG-111shrg58397/html/CHRG-111shrg58397.htm. Accessed August 1, 2013).

3

Strategic Planning and Metrics

This chapter addresses Task 1, which asks the committee to assess the feasibility of developing performance measures and quantitative metrics against existing performance goals of strategic plan and, if infeasible, recommended alternative approaches (see Appendix B for the full task statement).

This chapter comprises four sections. The first section introduces key terms, definitions, and concepts related to effectiveness measures and metrics. The second section assesses the feasibility of using measures and metrics to evaluate the effectiveness of the GNDA. The third section provides a summary of an analysis of the existing performance goals and strategic plan using the definitions and concepts introduced in the first two sections. The chapter concludes with a finding to address Task 1.

The following documents served as key references for this chapter:

- Government Performance and Results Act (GRPA) of 1993[1] and the GPRA Modernization Act of 2010[2]
- Office of Management and Budget (OMB) Circular No. A–11, Preparation, Submission, and Execution of the Budget (OMB, 2012)

---

[1] Pub. L. No. 103-62, § 20, 107 Stat. 285, available at http://www.whitehouse.gov/omb/mgmt-gpra/gplaw2m. Accessed August 1, 2013.

[2] Pub. L. No. 111-352, § 1, 124 Stat. 3866, available at http://www.gpo.gov/fdsys/pkg/BILLS-111hr2142enr/pdf/BILLS-111hr2142enr.pdf. Accessed August 1, 2013.

- Global Nuclear Detection Architecture Strategic Plan 2010 (GNDA, 2010*)
- Global Nuclear Detection Architecture Joint Annual Interagency Review (GNDA, 2011*; GNDA, 2012*)
- Department of Homeland Security (DHS) GNDA Implementation Plan 2012 (DHS, 2012*)

3.1 KEY TERMS AND DEFINITIONS

Multiple definitions can be found for the terms commonly used in performance measurement theory. The following terms and definitions are presented to clarify their use in this report for addressing this study's specific tasks.

Measure: Qualitative or quantitative facts that gauge the progress toward achieving a goal. These facts may be in the form of indicators, statistics, or metrics.

Indicator: A measurable value that is used to track progress toward a goal or target. (See "metric," below.) "Agencies are encouraged to use outcome indicators . . . where feasible" (OMB, 2012, Section 200, p. 14).

Metric: Synonymous with "indicator," the actual quantity that is used to measure progress. Metrics can be quantitative or qualitative. Quantitative metrics may use numerical (e.g., a percentage or number) or constructed (e.g., high, medium, low) scales.

Proxy metric: A metric that does not directly relate to a goal or objective but can be used as an indirect measure as long as a strong relationship exists between the metric and its objective can be made. Proxies can be useful and should not be indiscriminately avoided, especially when a direct metric cannot be established. Proxy metrics are also called "indirect metrics."

An example of a measure and its metric would be the percentage of planned portal monitors that have been deployed at seaports (measure) and the number of portal monitors deployed in the past year (metric). An example of a proxy metric is the number of preexisting memoranda of understanding (MOUs) for sharing equipment and resources between states that are established before a disaster. This proxy has been shown by

* Not publically available.

the Environmental Protection Agency to be directly related to how rapidly adjoining states can respond with additional equipment following disasters and emergencies (Travers, 2012).

The committee uses "metric" in place of "indicator" and simplifies "measures and metrics" to "metrics." At a detailed level, there is an important distinction between measures and metrics. However, the simplification allows this report to focus on the analysis and discussion of GNDA metrics rather than the differentiation between measures and metrics and to avoid the repeated use of the term "measures and metrics."

> **Goal:** A statement of the result or achievement toward which effort is directed. Strategic (or high-level) goals articulate clear statements of what the agency aims to achieve to advance its mission and address relevant national problem, needs, challenges, and opportunities. Such goals generally outcome-oriented and long-term and focus on major functions and operations of the agency.

> **Objectives:** Objectives directly link to a goal and reflect the outcome or impact the agency is trying to achieve.

> **Performance Goals:** Performance goals link to objectives and are established to help the agency monitor and understand progress. They should be of limited number and explain how they contribute to the strategic objective. "Agencies are strongly encouraged to set outcome-focused performance goals" (OMB, 2012, Section 200, p. 15).

This report does not distinguish between "strategic goals" and "goals" or "strategic objectives" and "objectives." The use of goals and objectives throughout this report assumes they are strategic (or high level).

> **Mission Statement:** "A brief, easy-to-understand narrative . . . [that] defines the basic purpose of the agency and is consistent with the agency's core programs and activities expressed within the broad context of national problems, needs, or challenges" (OMB, 2012, Section 200, p. 13).

> **Strategic Plan:** Presents the long-term objectives an agency hopes to accomplish. It describes general and longer-term goals the agency aims to achieve, what actions the agency will take to realize those goals, and how the agency will deal with the challenges likely to be barriers to achieving the desired result. An agency's strategic plan should provide the context for decisions about performance goals, priorities, and budget planning, and should provide the framework for the detail provided in agency annual plans and reports (OMB, 2012, Section 200, p. 13).

Types: There are different types of goals, objectives, and metrics (OMB, 2012, Section 200, page 14):

- Input—indicates consumption of resources used (e.g., time, money).
- Process—indicates how well a procedure, process or operation is working.
- Output—describes the level of activity (or product) that will be provided over a period of time.
- Outcome—indicates progress against achieving the intended result.

The committee introduces the concept of the functional scope of a goal, objective, or metric to describe how broadly focused it is.

Scope: The breadth of the focus of a goal, objective or metrics as they relate to the GNDA:

- Architecture—the integrated capability of all three geographic layers and the crosscutting functions of the GNDA;
- Layers—the operational elements and assets in each of the three geographical layers of the GNDA (exterior to the United States, transborder, and interior of the United States); and
- Resources/Assets/Capabilities—budgets, people, assets, and capabilities.

The remainder of this section discusses the characteristics and assessment of metrics that are useful to decision makers.

3.1.1 Characteristics of Metrics Useful to Decision Makers

Metrics are already used to report on the yearly progress of the GNDA. However, these metrics do not provide an assessment of the overall GNDA. This section introduces tools to develop and evaluate metrics in terms of their usefulness (e.g., their ability to provide information on the overall effectiveness of the GNDA).

Metrics are developed against particular criteria that are selected on the basis of the application (or objective) and the needs of the customer (or user). Different metrics may be chosen to meet different applications. For example, metrics could be used to assess U.S. security, report on management effectiveness, or gauge interagency cooperation. Customers who require reports on management effectiveness (e.g., Congress) may have a significantly different focus from customers interested in U.S. security assessment (e.g., GNDA federal partners). In fact, the GNDA has several customers for its metrics, for example:

	Type			
Scope	Input	Process	Output	Outcome
Architecture				
Layer				
Resources/ Assets/ Capabilities				

FIGURE 3-1 A scope-versus-type matrix is formed by combining the different metric types (input, process, output, or outcome) with their scope (full architecture level, layer, and resources). Similar matrices can be used to categorize goals and objectives.

- Congress and the White House (e.g., OMB or the National Security Council),
- Domestic Nuclear Detection Office (DNDO) and DHS management,
- GNDA federal partners, and
- Other GNDA partners (including foreign, state, and local jurisdictions).

Outcome-based metrics provide information that is useful to these customers because they provide information on progress made against an intended result or changes in conditions that the customer is attempting to influence. Broadly-based metrics that provide information on the full scope of the GNDA (i.e., the overall GNDA) will also be of more value to those customers than narrowly-scoped metrics. To guide the development of metrics that are both outcome-based and broadly-scoped, [3] a categorization matrix that includes the type of metric (e.g., input-, process-, or outcome-based) and its functional scope (e.g., architecture, layer, or resources) is introduced. This matrix can be used to categorize existing or proposed metrics (see Figure 3-1).

Goals, objectives, and metrics that populate the upper-right corner (outcome-based and focused on the full architecture) are preferred over those found in the lower-left corner (input-based and focused at the resource-level) of the matrix. Such matrices can aid in the development of updated goals, objectives and metrics that will inherently provide better

[3] There were some committee members who judged that this was redundant; if truly outcome-based, a metric (or goal or objective) would naturally be broadly-scoped. Because developing outcome-based strategic plans and metrics is difficult for programs such as the GNDA, it was determined that this additional criterion may prove helpful in GNDA agency self-assessment of future goals, objectives, and metrics.

information to decision makers than those that exist today. The development of outcome-based metrics that focus on the overall effectiveness of the GNDA requires that the higher-level goals and objectives also be outcome-based and focused on the full architecture. This is discussed in more detail later in this chapter.

Metrics that are useful tend to have the following additional characteristics:

- Understandable and transparent, including with respect to uncertainties—for confidence and communication and to enable peer review.
- Reproducible and flexible—to track progress and illustrate trends. When expert elicitation is used for data collection, assumptions and uncertainties should be provided.
- Quantitative with numerical or constructed (e.g., high/medium/low) scales.
- Verifiable—for credibility, quality control and confidence.

This list of characteristics is consistent with other formulations, such as those provided by Keeney and Gregory (2005)[4] and for software (Kaner, 2009). Checklists by themselves are not a guaranteed method of constructing meaningful or useful metrics. Foremost, metrics need to report on outcomes that are directly related to goals and objectives. Box 3-1 contains a summary of the characteristics of a useful metric.

---

[4] The characteristics listed in Keeney and Gregory (2005, p. 3) are listed below:

- Unambiguous—A clear relationship exists between consequences and descriptions of consequences using the attribute.
- Comprehensive—The attribute levels cover the range of possible consequences for the corresponding objective and value judgments implicit in the attribute are reasonable.
- Direct—The attribute levels directly describe to the consequences of interest.
- Operational—In practice, information to describe consequences can be obtained and value tradeoffs can reasonably be made.
- Understandable—Consequences and value tradeoffs made using the attribute can readily be understood and clearly communicated.

The list of criteria matches reasonably well with that proposed by the committee. The committee lists "understandable and transparent," which relates to "unambiguous" and to "understandable." "Reproducible and flexible" relates to "operational." The committee includes "quantitative," whereas Keeney and Gregory suggest that the attributes should be "direct" in describing the consequences of interest. "Verifiable" relates to "direct" and "operational." The concept of a "comprehensive," metric is implicit in the committee's discussion of the notion of "reproducible and flexible."

BOX 3-1
Characteristics of Metrics Useful to Decision Makers

Useful metrics have following characteristics:

1. Defined customers and an understanding of their applications
2. A clear connection to consequences or decision options for the customer by being:
   a. Outcome-based and broadly focused
   b. Aligned clearly to higher-level outcome-based goals and objectives
   c. Understandable/transparent, reproducible, quantifiable, and verifiable
   d. Directly linked to objectives and goals when output-, process- or input-based.

Items 2.a and 2.b are more critical characteristics of useful metrics than are 2.c and 2.d.

3.2 FEASIBILITY OF EVALUATING THE GNDA

The focus of Task 1 is to assess the feasibility of developing metrics against existing performance goals of the GNDA strategic plan that can evaluate the overall effectiveness of the GNDA. The remainder of this chapter focuses on the existing GNDA strategic plan and its performance goals and also on assessing the feasibility, using the concepts introduced in the first section, to develop metrics against them.

Section 1103 of the Implementing the Recommendations of the 9/11 Commission Act of 2007 (P.L. 110-53) mandates annual interagency reviews of the GNDA. The results of these annual reviews are submitted to the President, Congress, and OMB. Therefore, OMB's guidance on strategic planning and annual reporting are relevant to the GNDA and its federal partner agencies. Part 6 of Circular A-11 (OMB, 2012) describes the GPRA Modernization Act requirements and the expected approach to performance reporting.[5] OMB's strategic planning hierarchy can be found in Figure 3-2.

The guidance from OMB suggests that goals, objectives, performance goals, and metrics be *outcome oriented*, meaning that they should be focused on progress toward a mission rather than focused on activities and processes. Figure 3-2 shows the OMB "goals relationship" with some committee modifications: "Performance Goals with Performance Indicators"

[5] DNDO and its partner agencies released the GNDA strategic plan 2010; this guidance from OMB was released in 2012.

FIGURE 3-2 The Goals Relationship showing the hierarchical relationship between a single mission statement supported by multiple goals, which in turn are supported by a suite of objectives and performance goals and their associated metrics. The example goals and objectives are for illustration only. This diagram has been modified by the committee to reduce the number of terms that are used, bringing it in line with the text of this report and to draw parallels with the GNDA strategic plan. SOURCE: Modified from OMB (2012, Part 6, Section 200).

was reduced to simply "Performance Goals," and the examples within the "Other Indicators" were replaced with a subset of metric types introduced in the preceding section. The basic hierarchical structure remains the same.

In practice, metrics are created to report on progress in meeting the performance goals. The performance goals inform the progress in meeting the objectives which, in turn, inform progress toward meeting the goals. The goals link directly to the mission (Figure 3-2).

3.3 ANALYSIS

Is it suitable to develop additional metrics against the existing GNDA performance goals? The committee notes that it is exceedingly difficult to create outcome-based metrics for the GNDA when its higher-level strategic components (goals, objectives, performance goals) are not outcome-based and are not focused on full architecture. In determining whether a component is outcome-based, it is helpful to ask the question "Why is this being done?" or "Why is this important?" If the component is outcome-based, the answer will be self-evident. If an explanation of "Because . . . x, y, z," is needed to answer the question, then the component is not outcome-based.

The committee applied this test to the existing strategic plan. While the existing plan has a hierarchical structure similar to the OMB suggested structure (see Figure 3-2), the connection between the mission, goals and objectives was not clear. Furthermore, the committee determined that the majority of the existing goals and objectives are not outcome-based nor are they focused on the full architecture. Therefore, developing outcome-based metrics to report on progress of the overall GNDA against the existing plan is not feasible.

3.4 FINDING

FINDING 1.1: It is fundamentally possible to create outcome-based metrics for the GNDA; however, it is not currently feasible to develop outcome-based metrics against the existing strategic plan's goals, objectives, and performance goals because these components are primarily output- and process-based and are not linked directly to the GNDA's mission.

Two conditions must be met to use metrics to evaluate the effectiveness of the GNDA:

1. **A new strategic plan with outcome-based goals and objectives must be created and**
2. **An analysis framework must be developed to enable assessment of outcome-based metrics.**

The committee concludes that it is not suitable to develop further metrics against the current strategic plan because they would not be outcome-based.

Through the development of the strategic plan, DNDO and its partners have defined the GNDA from a large and complex set of disparate U.S. government programs. DNDO is using the annual review process as a mechanism to engage its partners in a cooperative effort to evaluate and improve the GNDA. In its present state, the strategic plan and annual review provide an accounting of the preexisting programs but not an assessment of overall performance. While acknowledging potential implementation challenges in Observations 1 and 2, the committee notes that further steps, including an updated strategic plan (see Chapter 4) and development of an analysis framework (see Chapter 5) are needed to transform this initial effort into an integrated analysis and planning capability that can better estimate overall GNDA effectiveness and inform decisions.

4

A Notional Strategic Planning Example with Metrics

The committee finds in the preceding chapter that it is fundamentally possible to measure the effectiveness of the Global Nuclear Detection Architecture (GNDA) as long as a new strategic plan and analysis framework are created. To demonstrate feasibility, a notional strategic planning example with outcome-based metrics is presented in this chapter and a new analysis framework is presented in Chapter 5. The committee found the development of the notional example to be difficult. This chapter will highlight the difficulties that the committee encountered (such as the development of deterrence metrics) and to provide its recommendations for the GNDA partner agencies if they decide to revise the GNDA strategic plan following the committee's recommendations. These recommendations are based both on the committee's expertise and the lessons learned from developing this notional example.

This chapter consists of three sections. In the first section, a notional example to address Task 1 is introduced. It shows how to structure a strategic plan to enable the development of outcome-based metrics. The second section presents a discussion on developing metrics against deterrence goals. This is intended to address the Domestic Nuclear Detection Office's (DNDO's) interest in deterrence and more broadly to highlight the challenges of developing metrics against preventive goals. The final section analyzes the notional example using criteria and tools introduced in Chapter 3. The chapter ends with recommendations for strategic planning and metrics development based on lessons learned from this exercise.

4.1 NOTIONAL EXAMPLE

There are significant challenges to developing metrics to demonstrate progress toward achieving the GNDA mission. OMB notes that the development of performance measures for "programs that relate to deterrence or prevention of specific behaviors" are particularly challenging (OMB, 2003, p. 10).

The following notional example of an outcome-based strategic plan with associated outcome-based metrics illustrates an approach (and an existence proof) for measuring GNDA effectiveness. This example shows how a strategic plan can be structured to allow for the development of outcome-based metrics.

A strategic plan lays out the vision, mission, and high-level goals and objectives of an organization or program. Additional details on how the objectives and goals can be implemented are frequently provided in an implementation plan. The vision statement describes the long-term goal(s) of the organization or program whereas the mission, coupled with the goals and objectives, describes how to achieve the vision through shorter-term goals. A vision statement need not be readily achievable or define a clear end state; for example, the Environmental Protection Agency's vision is "a cleaner, greener, more sustainable environment." Some say that a vision statement should grab the heart whereas the mission statement should speak to the mind.

Goals and their underlying objectives derive from the mission and are not limited to those that can be measured easily. Often, strategic planning is guided by available data rather than by determining a logical set of goals and objectives to address the mission.

In developing the notional example, the committee considered the broad challenge of protecting the nation from radiological and nuclear terrorist attacks and outlining a vision, a mission, and a set of goals and objectives that would directly support that broader challenge. By approaching the problem from this broad perspective, strategic planning is not affected by the limited, federally defined scope of "detection" (see Figure 2-1 and the committee's Observation 2).

The committee's notional example is illustrated schematically in Figure 4-1 and described in Box 4-1. It consists of the following elements:

- Vision: "For U.S. citizens to live free from the fear of nuclear or radiological terrorism."
- Mission: "Protect the nation from terrorist attacks that use radiological or nuclear (RN) materials."
- Several examples of outcome-based goals, objectives and metrics

Table 4-1 provides a list of the notional metrics. These metrics are meant to be a sampling of those that would be developed against a full strategic plan. In practice, each objective would have at least one associated metric. Some of these notional metrics were selected to highlight specific challenges. For example, the metric "Effectiveness of deterrence by denial" which directly supports the objective to "Deter terrorists' RN attacks by demonstrating high likelihood of failure," is discussed in detail in the Section 4.2.

Cost-effectiveness can be calculated using a combination of the metrics listed above but it is not identified as a separate metric. Within the proposed notional hierarchy in Figure 4-1, cost-effectiveness would be calculated by aggregating and comparing several different metrics. For example, a cost-effectiveness study might assess objectives related to threat, vulnerabilities, and consequences using the associated metrics and would weigh those results against other metrics (e.g., increased cost and side effects). Multiattribute utility analysis could be used to assign values to the different metrics (for more details on evaluating cost-effectiveness, see Appendix D).

Some notional objectives could be evaluated using proxy metrics. For example, the metric "Improve the probability of detecting an attempt to bring RN materials into the United States" could have proxy measures such as "Percent of POEs [Ports of Entry] with RN portals" or "number of interdictions of attempts to transport RN materials." To be useful for this purpose, however, proxies must have a direct relationship to the goals and objectives.

In the final section of this chapter, the committee evaluates the notional example metrics against the characteristics of "useful metrics" (see Box 3-1) and the "scope-versus-type" matrices. First, the committee considers the challenging case of developing metrics to measure deterrence.

4.2 METRICS TO MEASURE DETERRENCE

Although not the main focus of the committee's tasking, DNDO asked the committee to consider approaches for quantifying deterrence. The study of deterrence as it relates to the GNDA requires significantly more resources and time than was available for this study. However, the committee outlines several challenges associated with developing metrics to measure the effects of deterrence on the effectiveness of the GNDA. This discussion is not meant to overemphasize deterrence as a mission component to the GNDA—which is to both deter *and* detect.

Deterrence is a reduction in the likelihood that terrorists will attempt an attack using RN materials that are outside of regulatory control. This can be achieved by denial, increasing the cost and difficulty of an attack, providing a reward for not attacking, or retribution in case of a successful

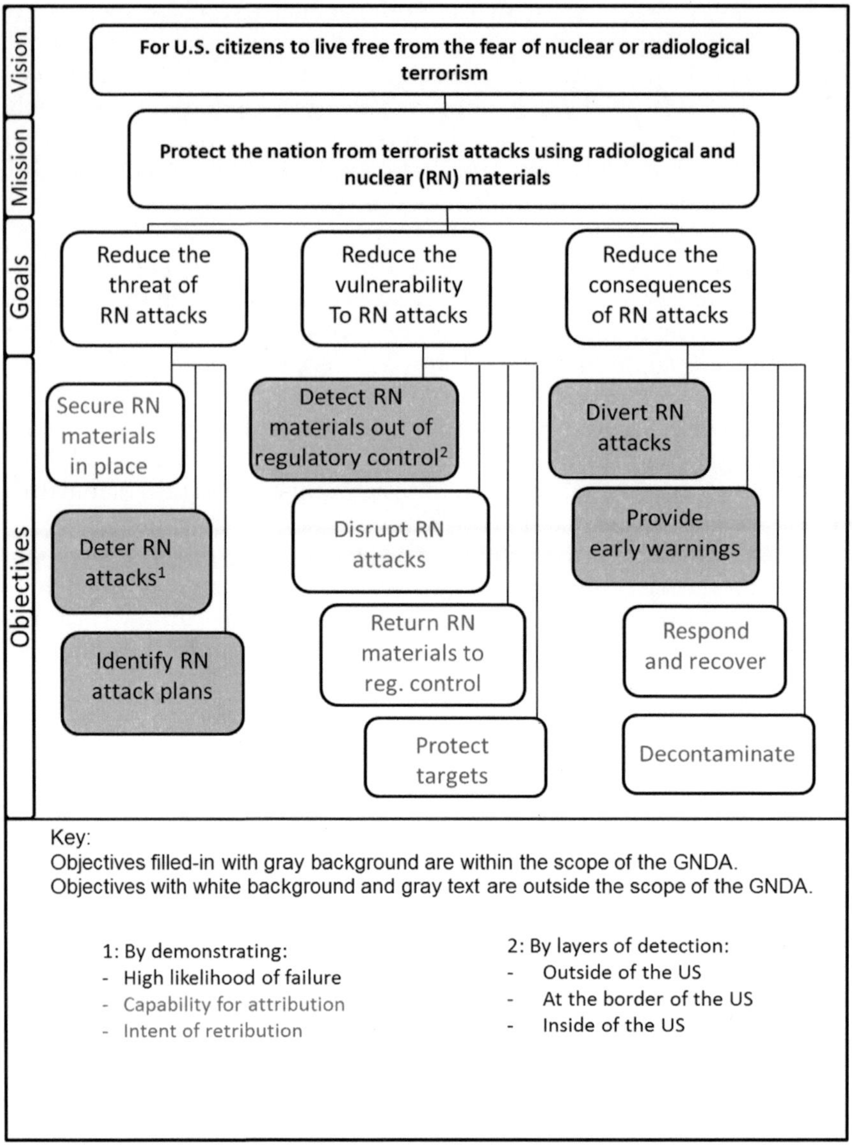

FIGURE 4-1 Notional example of a GNDA strategic plan outlined in Box 4-1. A vision, mission statement and a set of goals and objectives are presented. Text of the goals and objectives has been truncated to fit them into the diagram. Not all

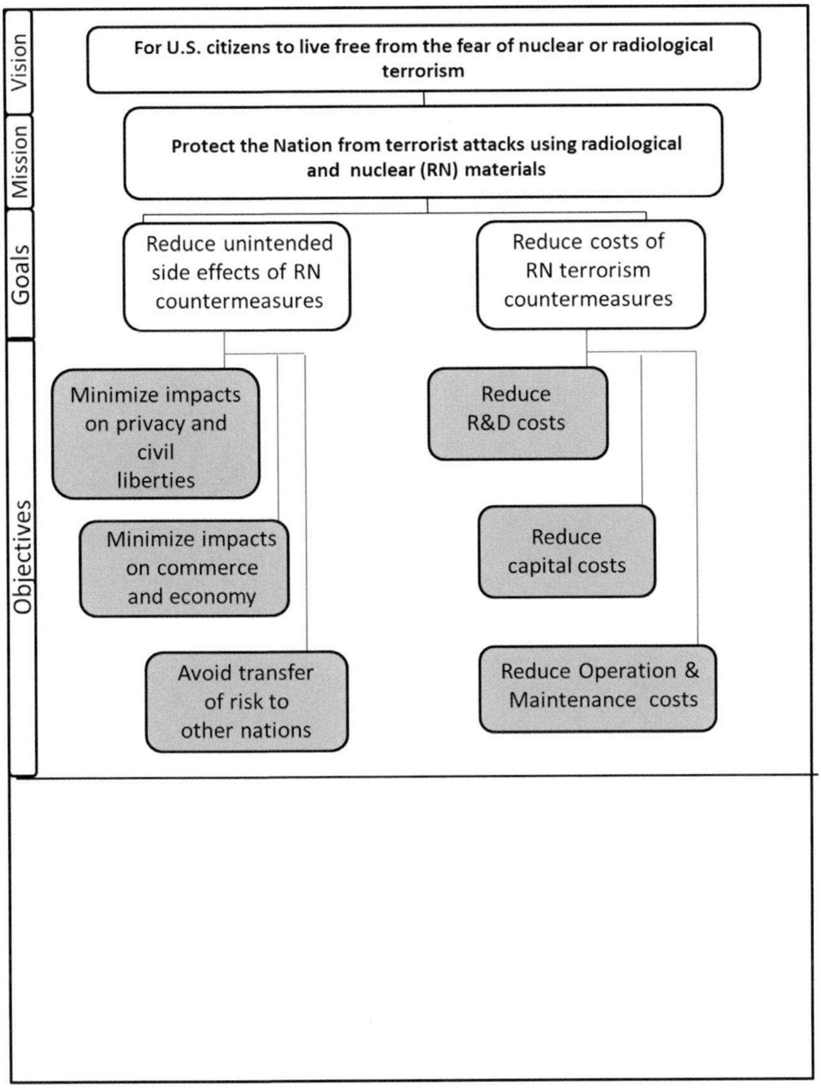

of the objectives are within the scope of the GNDA. Cells that have been filled in are within the scope of the GNDA; those not filled in are outside the scope of the GNDA.

BOX 4-1
Notional Strategic Plan with Metrics

Note: Objectives within the scope of the GNDA are highlighted by stars.

Vision: For U.S. Citizens to live free from the fear of nuclear or radiological terrorism
Mission: Protect the nation against terrorist attacks that use nuclear and radiological (RN) materials

Goal 1: Reduce the threat to the nation of radiological or nuclear (RN) attacks by terrorists.

- Objective: Secure RN materials in place.
- ★ Objective: Deter terrorists' RN attacks by demonstrating high likelihood of failure.
  - Metric: Effectiveness of deterrence by denial
- Objective: Deter terrorists' RN attacks by demonstrating the capability for attribution and intent of retribution.
- ★ Objective: Identify terrorist plans for radiological or nuclear attacks.

Goal 2: Reduce the vulnerability of the nation to RN attacks by terrorists.

- ★ Objective: Detect RN materials out of regulatory control by layers of detection:
  - outside of the United States,
  - at the border of the United States, and
    - Metric: Probability of detecting an attempt to bring RN materials into the United States at ports of entry (POEs)

attack. Deterrence by denial is achieved by demonstrating to the adversary that attacks would likely be detected, for example by hardening borders and restricting access to targets (NRC, 2002, p. 10). Efforts to increase the global effectiveness of detecting and reporting on RN material out of regulatory control contribute to deterrence by denial (e.g., increasing the cost to the terrorists by limiting the pathways available to move material without risk of detection), but they do not contribute to deterrence in other ways such as the threat of retribution. DNDO and GNDA partners are considering how to weigh efforts toward deterrence against the resulting threat reduction, which requires quantification.

The quantification and measurement of the impact of deterrence on terrorists has been discussed by several authors (Drake et al., 2003; Morral and Jackson, 2009; Willis et al., 2010; Haphuriwat et al., 2011). To assess the effectiveness of a deterrent, the following is needed at a minimum:

Metric: Probability of detecting an attempt to bring RN materials into the United States between POEs

inside of the United States

Objective: Disrupt terrorist attacks that use RN materials after detection

Objective: Return RN materials to regulatory control

Objective: Protect targets from RN attacks

Goal 3: Reduce the consequences of a successful radiological or nuclear attack on the nation

★ Objective: Divert RN attacks to lower-consequence targets

★ Objective: Provide early warning of RN attacks

Metric: Detection alert times

Objective: Respond and recover

Objective: Decontaminate

Goal 4: Reduce unintended side effects of RN countermeasures

★ Objective: Minimize impacts on privacy and civil liberties

★ Objective: Minimize impacts on the flow of commerce and the economy

★ Objective: Avoid transfer of RN risks to other nations

Goal 5: Reduce costs of RN countermeasures

★ Objective: Reduce research and development costs

★ Objective: Reduce capital costs

★ Objective: Reduce operations and maintenance costs

Metric: Total life-cycle cost

- Information about the objectives and values of the adversary—conceptually, the adversary's utility function for types of attack,
- Information about the adversary's perceptions of the likelihood of success of attacks,
- Information about the adversary's aversion to risk,
- Information about the adversary's decision rules when selecting attacks, and
- An accounting of the potential of shifting of risk from one target to another (displaced risk).

The challenges associated with gathering, quantifying, and measuring this information are discussed in the following paragraphs.

Measuring the impact of deterrence has been considered in other fields, for example, criminology (Anthony, 2004). *Deterrence and the Death Pen-*

TABLE 4-1 Set of Notional Metrics for the GNDA

GNDA Metrics (notional)
Effectiveness of deterrence by denial
Probability of detecting an attempt to bring RN materials into the United States at POEs
Probability of detecting an attempt to bring RN materials into the United States between POEs
Detection alert times
Total Life-cycle cost

NOTE: This set was developed to illustrate how outcome-based metrics might be developed against an outcome-oriented strategic plan. It is not a full set of metrics for the notional plan; in practice, metrics would be developed against the full set of objectives.

alty (NRC, 2012a) reviews the past 30 years of research on the impact of capital punishment on murder rates. The committee that authored this report notes the importance of understanding the perception of potential penalties by would-be murderers: "It is not possible to interpret empirical evidence of the relationship of homicide rates to sanctions without understanding how potential murderers perceive sanction regimes" (NRC, 2012a, p. 105). The challenge is that perception of risk—although a critical component of deterrence—is subjective and difficult to measure. One of the conclusions of the NRC report is that studies on perception need to take place. The report also notes the challenges of identifying the small subset of potential murderers within the broader population and, once identified, extracting truthful responses from them.

This concern is echoed in another study, *Discouraging Terrorism: Some Implications of 9/11* (NRC, 2002). In introducing the problem of measuring terrorists' perceptions of risk, the report notes that the impact of deterrence is difficult to measure when state actors are involved; it becomes significantly more difficult with terrorists. The challenges are identified as follows:

> (a) difficulties in getting unambiguous and credible threats across to terrorists,
> (b) the unwillingness of terrorists to communicate except indirectly and on their own terms,
> (c) exceptionally high levels of mutual distrust,
> (d) uncertainty about how to affect what terrorists value, and
> (e) uncertainty about the targets to which threats should be directed. (NRC, 2002, p. 1)

Measuring the impact of deterrence requires understanding and characterizing the decision-making process of would-be terrorists and what terrorists consider a successful attack (e.g., any RN device detonated at a target versus detonated upon interdiction). The report *Understanding and Managing Risk in Security Systems for the DOE Nuclear Weapons Complex* (NRC, 2011) concludes that there is not currently a comprehensive analytical basis to support assessment of the probability of adaptive adversaries' attacks.

Researchers have used game theory to model deterrence (Haphuriwat et al., 2011). Appendix D provides a simple model as an example to highlight the potential effectiveness of randomization strategies on adversaries' decisions. Lacking firm data on terrorists' perceptions, simple models can provide insight into potential strategies for deterrence. As with any modeling effort, validation through real events should be sought. Risk perceptions of adversaries will likely be collected and analyzed by intelligence agencies. Therefore, information from the intelligence community would be needed to address deterrence by denial through detection and reporting capabilities (e.g., What detection capabilities are known by potential terrorists? Does the physical location of detectors influence the attack strategies or plans of potential terrorists?).

The other aspect of measuring deterrence by denial is accounting for displaced risk (Morral and Jackson, 2009). The successful deterrence of a nuclear or radiological attack against a specific target needs to be weighed against the possibility that terrorists will then make alternative attack plans that could cause greater harm. Displacement can include shifts to other targets or other attack vectors (e.g., from nuclear to biological).

Deterrence is an important component of the GNDA mission even though it is difficult to characterize and measure. The committee did not find a credible or peer-reviewed deterrence model in its review. An indirect "deterrence effectiveness metric" could potentially be developed based on the assumption that the costlier an attack plan is for a terrorist organization, the less likely it is to happen. Other metrics identified in the notional example above could be combined and linked directly to increased costs of planning an attack and, as such, provide an indirect measure of the effectiveness of deterrence.

4.3 ANALYSIS OF NOTIONAL EXAMPLE

The committee provides an analysis of the notional example's strategic plan and metrics using the approach described in Chapter 3.

4.3.1 Customer and Application of Metrics

The application for the notional example was primarily performance based (for U.S. security). The customers would be GNDA partner agencies, Congress, and the White House. None of the notional metrics were developed purely for management; however, one such metric would be "the percent of existing nuclear detection capabilities currently integrated into a model."

4.3.2 Clear Connection to Consequences and Options

As discussed in Section 3.1, the most important criteria for metrics are that they be outcome-based and broadly focused. Figure 4-2 is the "scope-versus-type" matrix (introduced in Chapter 3) and is used here to analyze the notional metrics from Table 4-1.

This set of metrics is grouped toward the upper-right corner of the matrix which is the desirable region in this exercise. However, it is noteworthy that not all of the notional metrics are in the far upper-right corner box. In developing metrics, they will not always meet both criteria. Similar analyses can be conducted on the goals and objectives to illustrate that they are predominantly outcome-based and broadly-scoped. Consideration of the desirable characteristics of a metric (transparent, quantitative, reproducible, and verifiable) would show that not all of the notional metrics meet all of the characteristics. The matrices and the list of characteristics are tools that

	Notional Metrics Type			
Scope	**Input**	**Process**	**Output**	**Outcome**
Architecture		Detection alert times		Effectiveness of deterrence by denial Total lifecycle cost
Layer				Probability of detecting an attempt to bring RN materials into the U.S. at POEs Probability of detecting an attempt to bring RN materials into the U.S. between POEs
Resources/ Assets/ Capabilities				

FIGURE 4-2 A scope-versus-type matrix for the notional metrics from Table 4-1. The notional metrics have been categorized by type (input, process, output, and outcome) and scope (architecture, layer, resources/assets/capabilities).

may be used to guide future metric development—not all metrics will meet all of the criteria but striving to meet them will produce metrics that are more likely to be useful for assessing GNDA effectiveness.

4.4 RECOMMENDATION

RECOMMENDATION 1.1: When DNDO and the GNDA partner agencies next update the GNDA Strategic Plan, the committee recommends that they take the following steps:

1. **Generate a vision statement.**
   Without a clear, interagency-supported idea of the long-term goal of the GNDA, it is difficult to measure progress toward achieving it.
2. **Simplify the plan.**
   Limit the strategic plan's hierarchy to vision, mission, goals, and objectives; the goals and objectives should be outcome-based and they should clearly describe the desired results and how they are directly related to the mission and vision of the GNDA.
3. **Consider the broader nuclear counterterrorism problem before focusing on "detection."**
   A strategic plan developed by solely focusing on deterrence, detection, and reporting mission may not fully consider the activities that take place at the mission interfaces. Therefore, a broader perspective is needed to initially determine strategic goals and objectives before they are limited to those within the scope of the GNDA.
4. **Determine the goals and objectives by focusing on the mission.**
   Do not limit the plan's goals and objectives by focusing on what can be easily measured or by what data are by readily available. Some important objectives may not lend themselves to direct measurement but they should not be excluded from the plan for that reason.
5. **Use proxies when direct metrics are not available.**
   The metrics developed directly against outcome-based objectives will more readily be outcome-based and focused on measuring the full architecture. However, it is not always possible to develop metrics that meet these criteria. In those cases, proxies (i.e., indirect metrics which are frequently output- or process-based, such as the number of deployed detectors) can provide useful information as long as they can be directly linked to the objectives.

Furthermore, in the absence of a GNDA design document, the committee suggests that the strategic plan clearly describes the GNDA's design goals and how it enhances the otherwise disparate detection activities of GNDA partner agencies.

The findings and recommendations related to improved strategic planning efforts are general. They do not require or exclude specific types of planning (e.g., a capabilities-based planning). Therefore, the advice provided should be applicable regardless of the specific planning approach that the GNDA partners may decide to adopt should they proceed with implementation of the recommendations within this report.

Finally, the committee recognizes that significant organizational challenges exist that may impact the implementation of Recommendation 1.1—many of which are beyond DNDO's control. The committee did not investigate this topic further than the general observations made in Section 2.3.

5

GNDA Analysis Framework

As noted in Finding 1.1, two critical components are needed to evaluate the effectiveness of the Global Nuclear Detection Architecture (GNDA): a new strategic plan with outcome-based metrics and an analysis framework to enable assessment of outcome-based metrics. A notional example of a strategic plan and outcome-based metrics were developed in Chapter 4. The focus of this chapter is on an analysis framework. This chapter is intended to address the second charge of the statement of task (see Appendix B).

This chapter is organized into two sections. The first section provides an overview of current analysis approaches for complex technological systems. The final section introduces the committee's recommended GNDA analysis framework using the notional metrics described in the preceding chapter. The chapter concludes with two findings and one recommendation.

5.1 EVALUATION OF COMPLEX TECHNOLOGICAL SYSTEMS

The committee was asked to identify relevant examples of analytical risk-based approaches for complex technological systems. Military planning and analysis of defense capabilities provide several examples; water security provides another. These examples are described in this section.

Modeling plays an important role in defense analysis and support to decision makers. Military planning models exist at several levels: component, system, force structure (architectures), and campaigns. The models are used to inform force mission planning and force structure planning decisions.

For example, consider strategic airlift to support a military campaign. Different models have been created to support different planning decisions.

Component models exist of aircraft airframes to estimate drag and fuel consumption. System models exist to calculate the time to load an aircraft and deliver the cargo to a destination. Force structure models exist to calculate the time to deploy a force (people and equipment) for a mission. Finally, campaign models exist to determine the time to achieve the campaign objectives given the force available and potential actions of the adversary. For force structure planning, a variety of models are used to help determine the best mix of aircraft (e.g., tactical and strategic) and an affordable amount of airlift capability given the potential threats on the strategic planning horizon. The models do not make decisions; rather, they inform the analysts, strategic planners, and decision makers. They also analyze the advantages and disadvantages of viable alternatives.

Other organizations have developed analysis frameworks for complex systems that have addressed challenges similar to those faced by the GNDA architecture. These include U.S. strategic nuclear war planning, U.S. ballistic missile defense, and U.S. water security programs.

5.1.1 U.S. Strategic Nuclear War Planning

Strategic nuclear defense is an example of a complex technological system that uses an analysis framework (including modeling) to assess effectiveness. Since the beginning of the Cold War, the United States has relied on nuclear forces as a deterrent to hostile actions by nuclear adversaries. For obvious reasons, the United States cannot conduct a full- or even a partial-scale nuclear war to demonstrate the capability to deter adversaries. As a result, since the 1980s the United States has relied on a combination of modeling, war gaming, component testing, conventional system tests, reliability testing, red teaming, exercises, and technical studies to develop and evaluate the capabilities of its nuclear forces and to maintain a credible nuclear deterrent.

Nuclear force evaluation requires four types of data: adversary capabilities and intent, weapons availability, weapons reliability, and weapons effectiveness. Data on adversary capabilities and intent are obtained from expert elicitation of the intelligence community about adversary capabilities and possible intent. Weapons availability data are reasonably easy to obtain because U.S. military forces are required to collect these data. Reliability and effectiveness data are more difficult to obtain directly. Since the first nuclear test moratorium[1] took effect in 1992, the United States no longer fully tests the operation of nuclear warheads. Component, subsystem, and conventional (i.e., nonnuclear) system tests have been used to evaluate the

[1] The test ban began in 1992, introduced by the Hatfield-Exon-Mitchell amendment. The test ban moratorium continues to be upheld.

reliability of warhead material properties and control system electronics. In addition, conventional system tests have been used to evaluate the reliability of intercontinental missiles by selecting a missile at random, removing the warhead, shipping it to a test launch site, installing an electronic warhead simulator, and launching the missile to a downrange location in the Pacific. Extensive telemetry is collected during the tests to evaluate missile reliability. Based on the system reliability data obtained from these tests and evaluations, military planning factors have been developed to assess the availability, reliability, and probability of kill against potential adversary targets. Modeling and analysis have been used to develop and evaluate the overall effectiveness of the nuclear war plan and limited nuclear options, including the extensive Stockpile Stewardship Program that certifies nuclear arsenal readiness (NRC, 2012c). Modeling, analysis, and war gaming have also been used to consider and assess the potential actions of nuclear adversaries.

5.1.2 U.S. Ballistic Missile Defense

The U.S. National Missile Defense Act of 1999 (P.L. 106-38) states:

> It is the policy of the United States to deploy as soon as is technologically possible an effective National Missile Defense system capable of defending the territory of the United States against limited ballistic missile attack (whether accidental, unauthorized, or deliberate) with funding subject to the annual authorization of appropriations and the annual appropriation of funds for National Missile Defense.

Ballistic missile defense (BMD) is managed by the Missile Defense Agency (MDA) in the Department of Defense. The system's architecture includes:[2]

- networked sensors (including space-based) and ground- and sea-based radars for target detection and tracking;
- ground- and sea-based interceptor missiles for destroying a ballistic missile using either the force of a direct collision, called "hit-to-kill" technology, or an explosive blast fragmentation warhead;
- and a command, control, battle management, and communications network providing the operational commanders with the needed links between the sensors and interceptor missiles.

Like the GNDA, the BMD is a complex detection architecture composed of a system of systems to defend against intelligent, adaptive adversaries for a high-consequence event that has not yet occurred. And like

[2] See http://www.mda.mil/system/system.html. Accessed August 1, 2013.

the GNDA, the BMD cannot be evaluated by the direct use of operational stimuli (Parnell et al., 2001; Garrett et al., 2011; Willis, 2012). BMD also has a critical time challenge for reporting and response due to the short flight times of ballistic missiles. The BMD mission also has a critical deterrence component: The United States wants to deter an adversary from attacking with nuclear ballistic missiles.

The major challenges for analyzing the effectiveness of the BMD architecture involve characterizing adversary objectives, the operational performance of the adversary ballistic missiles and warheads, and the performance of the U.S. BMD architecture. Choices for an optimized command-and-control strategy for allocating defensive assets against adversary offensive assets can then be determined. Again, there is no way to obtain a full operational evaluation of the BMD architecture. One must rely on intelligence data to obtain adversary objectives and testing to provide missile and weapon performance data. For U.S. BMD systems, military planning factors include the assessment of the availability, reliability, and probability of kill against potential BMD targets based on component, subsystem, and limited-engagement ballistic missile test data (one defense system versus one simulated adversary missile). Modeling, analysis, exercises, and red teams are used to develop and evaluate the overall effectiveness of the BMD architecture. Because adversaries will seek to achieve their objectives by exploiting BMD vulnerabilities, these models must explore the full range of potential adversary objectives and attack plans. Again, one must evaluate the full architecture because the improvement of the defense in one region may shift the risk to another region.

5.1.3 U.S. Water Security Program

The Environmental Protection Agency's (EPA's) Water Security Office works with the Department of Homeland Security (DHS) to protect the U.S. water and wastewater critical infrastructure against all hazards (Travers, 2012). The Water Security Office shares many of the same challenges as the GNDA: it has a complex architecture; it must defend against adaptive adversaries; and the EPA's Water Security Office lacks regulatory and budgetary authority over national, industrial, state, and local stakeholders. There are approximately 70,000 industrially owned water and wastewater treatment facilities throughout the United States of varying sizes and complexity. In addition to the risk of contamination by a terrorist, environmental hazards must also be included in the evaluation of overall risk. In the United States, major contamination events have not occurred and environmental disasters affecting the water supply are extremely rare; consequently, operational stimuli are rare or nonexistent. Unlike other parts of the EPA, the Water Security Office does not have regulatory authority to enforce security practices of industrial, state, and local stakeholders.

The Water Security Office relies on a combination of models, simulations, and exercises to evaluate the effectiveness of water security programs. The models and simulations are exercised at state and local levels. The results of exercises are used to increase fidelity of the models. Incentives are used to encourage data sharing and involvement of stakeholders. These include developing strong partnerships with associations to gain local trust and establish legitimacy, developing products that all stakeholders can use (e.g., a downloadable, do-it-yourself security risk assessment tool), communicating the importance of managing risk for rare-occurrence, high-consequence events, and highlighting the mutual benefits of implementing water security programs (e.g., contamination detection systems also detect early-stage corrosion problems, which if corrected could allow for cost-effective solutions). The Water Security Office strives to use unclassified and unrestricted information to increase communication and use of products by stakeholders.[3] The Water Security Office also recognizes the importance of deterrence through strong security but does not tie performance objectives to this measure.

5.2 PROPOSED GNDA ANALYSIS FRAMEWORK

The committee's proposed GNDA Analysis Framework provides a process for evaluating the notional metrics of the GNDA strategic plan. The framework provides an analytic capability to evaluate GNDA effectiveness and inform cost-effective resource decisions. It is described using the notional strategic plan and metrics examples presented in Chapter 4. Figure 5-1 provides a description of this framework. At the top of this figure are strategic planning, potential resource decisions that need to be made, and decisions that have been made based on analyses of the framework's metrics. The GNDA Architecture Definition (left side of Figure 5-1) includes conceptual data (e.g., GNDA resources within the layers and crosscutting functions) as well as the physical information (e.g., locations of existing assets, detector types and detector performance, established communications channels, and analysis capabilities). Models include risk models and probabilistic network models but incorporate other models as shown. Metrics (right side of the figure; e.g., notional metrics from Chapter 4) are generated from a set of models for a variety of customers; data for metrics may also be generated from other means (e.g., timelines and budgets). Validation and Verification (V&V; bottom of the figure) is needed for a credible analysis

[3] A copy of the memorandum from Michael H. Shapiro is available on the EPA website at http://water.epa.gov/infrastructure/watersecurity/lawsregs/upload/policytomanageaccesstosensitivedwrelatedinfoApril2005.pdf.

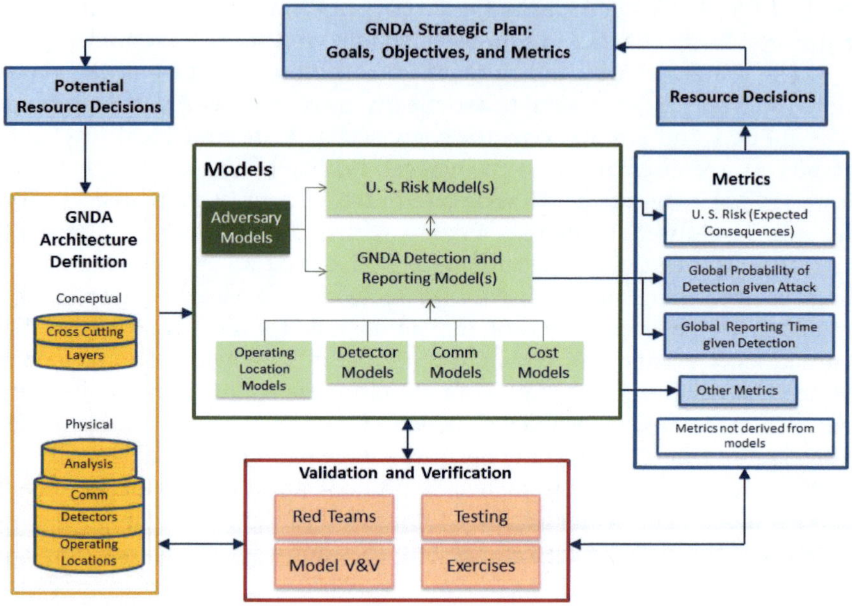

FIGURE 5-1 Proposed GNDA Analysis Framework. This framework has four major components with critical linkages to strategic planning: GNDA Architecture Definition (resource information such as detectors, personnel), Models, Metrics, and Validation and Verification (V&V). This framework guides strategic planning and allows for prioritization of goals and objectives by assessing the impacts of alternative resource allocations.

framework (Figure 5-1). V&V activities can also be used to determine or confirm uncertainties.

5.2.1 Alignment with Strategic Planning and Resource Decisions

One of the key purposes of an analysis framework is to provide iterative assessments to inform the GNDA strategic planning process. Strategic planning is not part of the analysis framework per se but it is connected to it in two ways: the framework can be used to evaluate and prioritize potential resource decisions, and the metrics' output from the models can be used to demonstrate the effectiveness of previously made resource decisions. Potential resource decisions (e.g., new detectors, expanded training, and changes in budgets) are reflected in the GNDA Architecture Definition (left side of Figure 5-1). GNDA resource decisions also include operational changes (e.g., potential detector resource randomization strategies). The analysis framework evaluates the architecture's proposed design using metrics; the resulting change in metrics is used to guide resource decisions on potential improvements to meet objectives and goals. The analytical process is iterative and can therefore be used to explore a large number of potential resource decisions. Once a resource decision has been made and implemented, the framework's metrics evaluate progress made toward meeting the strategic plan's goals and objectives.

5.2.2 GNDA Architecture Definition

The GNDA Architecture Definition is a description of a set of detection and reporting capabilities (or resource allocations) arranged against a specific threat. The definition can be thought of as input data for the models. The purposes of the analysis framework are to evaluate the architecture's integrated design, assess potential improvements, and to identify cost-effective improvements to meet strategic objectives. The GNDA architecture must be defined at the conceptual and physical data levels. There must be a common understanding and definition of existing GNDA architecture data (e.g., detectors, communications links, trained people) to calculate the metrics required for the assessments.

5.2.3 Models

Models are the key components of the proposed analysis framework illustrated in Figure 5-1. The reason for this is simple: as described in Section 5.1, models are essential for understanding the performance of complex systems (see also Appendix D).

Models have several important uses. They integrate data from a wide variety of GNDA-related programs and are used to exercise the GNDA against a variety of adversary stimuli. Models are also used to calculate the metrics based on other architecture data (such as data from tests, historical records, or expert elicitations). The models can be used to evaluate the current architecture, potential improvements, and their costs. Models such as RNTRA and PEM can be used in the committee's proposed analysis framework. In addition, other models such as adaptive adversary models (see Figure 5-1) will also be needed.

The models may not capture all of the components of existing GNDA system. In fact, it may not be cost-effective to model all of the resources or programs that currently are listed in the annual reviews. However, the target percentage of GNDA representation should be a conscious decision made by decision makers with input from key stakeholders. DNDO could consider dynamically linking its models to real time data bases maintained in its analysis centers (such as the Joint Analysis Center, JAC, and others) to ensure that these models reflect the architecture as deployed.

5.2.4 Metrics

Metrics are intended to provide information on progress toward meeting objectives and goals of the GNDA strategic plan and evaluating GNDA effectiveness. The framework provides an analytical capability to produce metrics for these purposes. GNDA effectiveness can be characterized using the set of notional GNDA metrics that were introduced in Chapter 4 (see Table 4-1) and other metrics listed in Figure 5-1:

1. **U.S. Risk (Expected Consequences).** Although not a GNDA responsibility, the proposed analysis framework needs to provide data to support risk analyses. This metric is not listed in Table 4-1 because it does not directly link to the objectives in the notional strategic plan.
2. **Effectiveness of Deterrence.** As discussed in Chapter 4, measurement of deterrence will be indirect. One could develop a proxy metric by assuming that an increased cost to the adversary will be incurred by increased detection capabilities and reduced detection times.
3. **Probability of Detecting an Attempt to Bring RN Materials into the United States at or between Ports of Entry.** To fully evaluate the GNDA, the architecture must be evaluated against a large number of potential attack plans to identify weaknesses that an adaptive adversary may try to exploit. This is a fundamental effectiveness measure.
4. **Detection Alert Time.** Threat detections are necessary but not sufficient for assessing GNDA effectiveness. The sensor or sensor operator needs to send threat detection information to key operational command-and-control centers in time for decision makers to analyze, determine, and authorize appropriate actions. Alert times can be modeled in a network model.

As discussed in Chapter 4, in practice each objective would have at least one metric developed against it.

5.2.5 Validation and Verification Activities

V&V play a critical role in the proposed analysis framework: validation ensures that the GNDA models capture the architecture and detection capabilities as they exist and verification confirms that the model is correctly performing the calculations that it claims to be performing. Independent V&V activities are needed to increase the fidelity of and confidence in the models, architecture data, and metrics.[4]

The following is needed for GNDA V&V:

- A complete technical description of the model and data (NRC 2008);
- Architecture designs, models and metrics validated by stakeholders to capture existing capabilities and proposed changes; and
- Credible peer review of the model.

Although the GNDA architecture design, models and metrics cannot be operationally evaluated (similar to BMD and strategic nuclear defense), specific parts can be validated using operational and developmental test data, red team assessments, exercises, and pilot activities. Several current examples of these types of activities could be used for validation.

DNDO's Red Team and Net Assessments (Oliphant, 2012) and the Operations Support Directorates (OSD) (Fisher, 2012) train, exercise and evaluate specific components of the domestic portion of the GNDA.

Other GNDA stakeholders participate in large-scale exercises such as National Level Exercises (NLEs), Alpha Omega or Marble Challenge 2010 and National Nuclear Security Administration's (NNSA) exercises that secure RDD threats.[5] Such exercises are performed regularly to demonstrate federal, state, and local coordination capabilities. Data from these exercises can be incorporated into the analysis framework to validate models and data. The committee notes that some of the nuclear counter-terrorism activities that these exercises evaluate may fall outside of the scope of the GNDA (see Observation 2), but that there is still benefit in these exercises for state and local communication and coordination and standard operating procedures (SOP) development.

Operational data can be used for V&V purposes as well. An example of non-domestic operational data that could be used for validation is Second Line of Defense programs (Leffer, 2012). The committee judges that these activities that could serve to validate an analysis framework do not currently incorporate their results into the GNDA architecture definitions, models, or metrics.

---

[4] For example, the Transportation Security Administration recently had an independent assessment performed on its analysis tools (Morral et al. 2012).

[5] See http://nnsa.energy.gov/mediaroom/pressreleases/bearcatexercise080912. Press release from the DOE's National Nuclear Security Administration. Accessed April 29, 2013.

5.4 FINDINGS AND RECOMMENDATIONS

The committee produced two findings and one recommendation in response to the second charge of the study.

FINDING 2.1:

A new GNDA Analysis Framework is needed to assess the GNDA effectiveness as shown in Figure 5-1. The critical components of the framework are the following:

1. **A GNDA Strategic Plan that contains outcome-oriented, broadly-scoped goals, objectives, and metrics and is directly connected to the components listed below;**
2. **A GNDA Architectural Definition that provides the conceptual and physical descriptions of the GNDA, and that define needed input data for the models described below;**
3. **A suite of GNDA models that incorporate potential adversary objectives, accurately represent existing and potential architecture capabilities, and calculate the metrics described below;**
4. **Metrics that can gauge overall GNDA effectiveness and assess potential GNDA resource decisions to increase GNDA effectiveness; and**
5. **A Validation and Verification (V&V) program that evaluates the data used in the GNDA architecture definition, models. and metrics. A robust V&V program enhances the credibility of the analysis framework.**

FINDING 2.2:

Current DNDO modeling, testing, red teaming, analysis, and training capabilities provide a foundation for evaluating components of the GNDA, but these current capabilities are insufficient for validating and verifying the overall effectiveness of the GNDA. Evaluating the effectiveness of the overall GNDA requires an integrated and continuous model-based scenario testing, red teaming, analysis, peer review, and training supplemented with intelligence awareness.

RECOMMENDATION 2.1:

DNDO should develop a new GNDA Analysis Framework similar to the framework proposed by the committee. This framework defines an analytic process that clarifies the connections among strategic planning, architectural definition, models, metrics, and validation and verification efforts. Such an analysis framework can provide credible assessments of overall GNDA effectiveness.

A new analysis framework can also provide a credible, transparent, and actionable GNDA cost-benefit modeling capability to support resource allocation decision making within GNDA federal agencies and collaborating nations that support the GNDA.

References

Anthony, R. 2004. A calibrated model of the psychology of deterrence. *Bulletin on Narcotics* 56(1/2):49-64.

Baldini, E. 2010. Preventing the theft of dangerous radiological materials. *The Police Chief* 77:30-39. Available at http://www.policechiefmagazine.org/magazine/index.cfm?fuseaction=display_arch&article_id=2199&issue_id=92010. Accessed August 1, 2013.

Cuellar, L., D. Kubicek, P. Stroud, M. Mathis, J. Smith, and F. Roach. 2012. Probabilistic Effectiveness Model, Data Documentation v1.2. LA-CP-12-00659. Los Alamos National Laboratory. May 3.

DHS (Department of Homeland Security). 2006. National Planning Scenarios Created for Use in National, Federal, State, and Local Homeland Security Preparedness Activities, Ver. 21.3. Available at https://www.llis.dhs.gov/content/national-planning-scenarios-final-version-213-0. Accessed August 1, 2013.

DHS. 2007. DHS' Domestic Nuclear Detection Office Progress in Integrating Detection Capabilities and Response Protocols. DHS Office of Inspector General, OIG-08-19.

DHS. 2011a. 2011 Radiological/Nuclear Terrorism Risk Assessment, Executive Summary. (SECRET)

DHS. 2011b. 2011 Radiological/Nuclear Terrorism Risk Assessment Final Report, Volumes I, II, III. (SECRET)

DHS. 2012. The Department of Homeland Security GNDA Implementation Plan. Report to Congress Fiscal Year 2012. (OUO)

Drake, G., W. Paddon, and D. Ciechanowski. 2003. Can We Deter Terrorists from Employing Weapons of Mass Destruction on the U.S. Homeland? Carlise, PA: U.S. Army War College.

Fisher, J. J. 2012. Training, Assistance, and Exercises Overview. Presentation to the Committee on Evaluating the Effectiveness of the Global Nuclear Detection Architecture, October 10, Washington DC. (OUO)

Garrett, R. K. Jr., S. Anderson, N. T. Baron, and J. D. Moreland, Jr. 2011. Managing the interstitials, a system of systems framework suited for the ballistic missile defense system. *Systems Engineering* 14(1):87-109.

GAO. 2008. Preliminary Observations on the Domestic Nuclear Detection Office's Efforts to Develop a Global Nuclear Detection Architecture. GAO-08-999T. July. Available at http://www.gao.gov/products/GAO-08-999T. Accessed August 1, 2013.

GAO. 2012. DHS Has Developed Plans for Its Global Nuclear Detection Architecture, but Challenges Remain in Deploying Equipment. GAO-12-941T. July. Available at http://www.gao.gov/products/GAO-12-941T. Accessed August 1, 2013.

Global Nuclear Detection Architecture (GNDA). 2010. The GNDA Strategic Plan. (OUO)

GNDA. 2011a. Joint Annual Interagency Review from 2011. (OUO)

GNDA. 2011b. Joint Annual Interagency Review of the Global Nuclear Detection Architecture Data Call for the 2011 Report. (OUO)

GNDA. 2012a. Joint Annual Interagency Review from 2012. (OUO)

GNDA. 2012b. Joint Annual Interagency Review of the Global Nuclear Detection Architecture Data Call for the 2011 Report. (OUO)

Haphuriwat, N., V. M. Bier, and H. H. Willis 2011. Deterring the smuggling of nuclear weapons in container freight through detection and retaliation. *Decision Analysis* 8(2):88-102.

Jain, M., J. Tsai, J. Pita, C. Kiekintveld, S. Rathi, M. Tambe, and F. Ordonez. 2010. Software assistants for randomized patrol planning for the LAX airport police and the Federal Air Marshal Service. *Interfaces* 40:267-290.

JASON. 2012. Domestic Nuclear Detection Surge Operations, JSR-11-315, 2012. (OUO)

Kaner, C. 2009. Metrics, Qualitative Measurement, and Stakeholder Value, Conference of the Association for Software Testing, July 2009.

Keeney, R. L., and R. S. Gregory. 2005. Selecting attributes to measure the achievement of objectives. *Operations Research* 53(1):1-11.

Leffer, T. N. 2012. Implementation of the Global Nuclear Detection Architecture, Part II: Interagency Partnership's Perspectives. Presentation to the committee, May 15, Washington, DC. (OUO)

Morral, A. R., and B. A. Jackson. 2009. Understanding the Role of Deterrence in Counterterrorism Security. Santa Monica, CA: RAND Corp. Available at http://www.rand.org/pubs/occasional_papers/OP281. Accessed August 1, 2013.

Morral, A.R., C. C. Price, D. S. Ortiz, B. Wilson, T. LaTourrette, B. W. Mobley, S. McKay, and H. H. Willis. 2012. Modeling Terrorist Risk for the Air Transportation System. Santa Monica, CA: RAND Corp. Available at http://www.rand.org/content/dam/rand/pubs/monographs/2012/RAND_MG1241.pdf. Accessed August 1, 2013.

NAE/NRC (National Academy of Engineering and National Research Council). 2004. Radiological Attack: Dirty Bombs and Other Devices. Report Brief. Available at http://www.dhs.gov/xlibrary/assets/prep_radiological_fact_sheet.pdf. Accessed August 1, 2013.

NAE/NRC. 2005. Nuclear Attack. Report Brief. Available at http://www.dhs.gov/xlibrary/assets/prep_nuclear_fact_sheet.pdf. Accessed August 1, 2013.

NEHRP (National Earthquake Hazards Reduction Program). 2008. Strategic Plan for the National Earthquake Hazards Reduction Program Fiscal Years 2009-2013. Available at http://www.nehrp.gov/pdf/strategic_plan_2008.pdf. Accessed September 5, 2013.

NNSA (National Nuclear Security Administration). 2012a. NNSA conducts 100th WMD counterterrorism exercise. Press Release. Available at http://nnsa.energy.gov/mediaroom/pressreleases/bearcatexercise080912. Accessed August 1, 2013.

NNSA. 2012b. NNSA sees significant achievements, important improvements in 2012. Press Release. Available at http://www.nnsa.energy.gov/mediaroom/pressreleases/achievements2012. Accessed August 1, 2013.

NRC (National Research Council). 2002. Discouraging Terrorism: Some Implications of 9/11. Washington, DC: The National Academies Press. Available at http://www.nap.edu/catalog.php?record_id=10489. Accessed September 9, 2013.

NRC. 2008. Department of Homeland Security Bioterrorism Risk Assessment: A Call for Change. Washington, DC: The National Academies Press. Available at http://www.nap.edu/catalog.php?record_id=12206. Accessed September 9, 2013.

NRC. 2010. Review of the Department of Homeland Security's Approach to Risk Analysis. Washington, DC: The National Academies Press. Available at http://www.nap.edu/catalog.php?record_id=12972. Accessed September 9, 2013.

NRC. 2011. Understanding and Managing Risk in Security Systems for the DOE Nuclear Weapons Complex, Abbreviated Version. Washington, DC: The National Academies Press. Available at http://www.nap.edu/download.php?record_id=13108. Accessed September 16, 2013.

NRC. 2012a. Deterrence and the Death Penalty. Washington, DC: The National Academies Press. Available at http://www.nap.edu/catalog.php?record_id=13363. Accessed September 9, 2013.

NRC. 2012b. Improving Metrics for the Department of Defense Cooperative Threat Reduction Program. Washington, DC: The National Academies Press. Available at http://www.nap.edu/catalog.php?record_id=13289. Accessed September 9, 2013.

NRC. 2012c. The Comprehensive Nuclear Test Ban Treaty: Technical Issues for the United States. Washington, DC: The National Academies Press. Available at http://www.nap.edu/catalog.php?record_id=12849. Accessed September 9, 2013.

NTHMP (National Tsunami Hazard Mitigation Program). 2013. National Tsunami Hazard Mitigation Program 2009-2013 Strategic Plan. Available at http://nthmp.tsunami.gov/documents/nthmp_strategicplan.doc. Accessed August 1, 2013.

Oliphant, M. 2012. Red Teaming Overview. Presentation to the committee, October 10, Washington, DC. (OUO)

OMB (Office of Budget and Management). 2003. Performance Measurement Challenges and Strategies. Available at http://www.whitehouse.gov/sites/default/files/omb/part/challenges_strategies.pdf. Accessed September 5, 2013.

OMB. 2012. Preparation, Submission, and Execution of the Budget. Available at http://www.whitehouse.gov/sites/default/files/omb/assets/a11_current_year/a_11_2012.pdf. Accessed August 1, 2013.

Parnell, G., E. Ronald, R. Metzger, J. Merrick, and R. Eilers. 2001. Multiobjective decision analysis of theater missile defense architectures. *Systems Engineering* 4(1):24-34.

Rosoff, H., and D. von Winterfeldt. 2007. A risk and economic analysis of dirty bomb attacks on the Ports of Los Angeles and Long Beach. *Risk Analysis* 27(3):533-546.

Shea, D. 2008. The Global Nuclear Detection Architecture: Are We Building Domestic Defenses That Will Make the Nation Safer? Testimony to the Committee on Homeland Security and Governmental Affairs. U.S. Congress, Senate.

Smith, J. 2012. Probabilistic Effectiveness Methodology (PEM). Presentation to the Committee on Evaluating the Effectiveness of the Global Nuclear Detection Architecture, October 10, Washington, DC.

Streetman, S. 2012. Radiological and Nuclear Terrorism Risk Analysis (RNTRA): DNDO's Risk Model. Presentation to the Committee on Evaluating the Effectiveness of the Global Nuclear Detection Architecture, June 28, Washington, DC. (OUO)

Travers, D. 2012. Developing and Using Metrics for the Water Security Program. Presentation to Committee on Evaluating the Effectiveness of the Global Nuclear Detection Architecture, October 10, Washington, DC.

U.S. Congress, Senate. 2010. Nuclear terrorism: strengthening our domestic defenses. S. Hrg. 1096 before the Committee on Homeland Security and Governmental Affairs, 111th Cong., 2nd Sess. Available at http://www.gpo.gov/fdsys/pkg/CHRG-111shrg58397/pdf/CHRG-111shrg58397.pdf. Accessed August 1, 2013.

Willis, H. H. 2012. Overview of Recent Workshop on Connecting Analysis to Strategic Planning with Examples. Presentation to the committee. June 28. Washington, DC.

Willis, H. H., J. B. Predd, P. K. Davis, and W. P. Brown. 2010. Measuring the Effectiveness of Border Security Between Ports-Of-Entry. Santa Monica, CA: RAND Corp. Available at http://www.rand.org/pubs/technical_reports/TR837.

Wyss, J.C. 2012. State Department's Role in GNDA, International Implementation Plan. Presentation to committee. October 10. Washington, DC. (OUO)

Zabko, J. 2012a. DNDO comparison between the GNDA's capabilities and expected adversary capabilities. Presentation to committee October 12. Washington, DC. (SECRET)

Zabko, J. 2012b. The global nuclear detection architecture: Metrics and evaluation. Presentation to committee May 14. Washington, DC. (OUO)

Appendix A

Presentations and Committee Information Gathering Meetings

Washington, DC, May 14-15, 2012

- *Study Background, Motivation and Challenge to the Committee,* Publicly releasable, Brendan Plapp, Department of Homeland Security, Domestic Nuclear Detection Office, Architecture and Plans Directorate
- *What Were the Original Intentions for the GNDA?,* Major General Julie A. Bentz, National Security Council
- *Description of Existing GNDA Structure and GNDA Strategic Plan,* Brendan Plapp, DHS/DNDO Architecture and Plans Directorate
- *DHS Global Nuclear Detection Architecture Implementation Plan Performance Measures,* Kevin Hart, DHS/DNDO Architecture and Plans Directorate
- *The Global Nuclear Detection Architecture: Metrics and Evaluation,* John Zabko, DHS/DNDO Architecture and Plans Directorate
- *Implementation of the Global Nuclear Detection Architecture, Part II: Interagency Partnership's Perspectives,* Teri N. Leffer, Department of Energy, National Nuclear Security Administration Second Line of Defense (NA-256)
- *Historical Perspectives and Congressional Authorities,* Dana Shea, Congressional Research Service (CRS)

Washington, DC, June 28-29, 2012

- *DNDO's Risk Model*, Steven Streetman, Data Architecture Solutions, Inc., supporting DHS/DNDO Architecture and Plans Directorate
- *Overview of Recent Workshop on Connecting Analysis to Strategic Planning with Examples*, Henry Willis, RAND
- *Global Perspectives and Activities of the GNDA*, David Kulp, Department of Defense
- *DNDO Operations Support Directorate: Domestic Implementation of GNDA*, Ernest Muenchau, retired Assistant Director OSD (deceased)
- *GNDA Concept of Operations Template*, Colonel Robert Kolterman, DTRA
- *Joint Analysis Center Overview*, Brian Savage, DNDO OSD

Long Beach, CA, August 28-29, 2012

- *Joint Regional Intelligence Center tour*
- *Panel Discussion*, U.S. Customs and Border Patrol, U.S. Coast Guard, Los Angeles Police Department, Federal Bureau of Investigation, L.A. Port Police, and L.A. Sheriff
- *Port Briefing*, Captain John Holmes, Port of Los Angeles
- *Port of Long Beach, CBP tour*
- *National Marine Exchange, Los Angeles*

Washington, DC, October 10-12, 2012

- *Risk Analysis & Decision Support,* James Smith, Los Alamos National Laboratory
- *Red Teaming Overview,* Mark Oliphant, DNDO, Red Team and Net Assessments
- *DHS Program Assistance: Training and Exercises as Ways to Measure the Effectiveness of the GNDA,* J. J. Fisher, DNDO, Operations Support Directorate
- *FBI Role in the Global Nuclear Detection Architecture (GNDA),* Bernie Bogdan, FBI
- *State Department's Role in GNDA: International Implementation Plan,* J. C. Wyss, State Department
- *Developing and Using Metrics for the Water Security Program,* David Travers, U.S. Environmental Protection Agency

Appendix B

Statement of Task

An ad hoc committee will conduct a study and prepare a report to the Domestic Nuclear Detection Office (DNDO) on quantitative approaches for evaluating the effectiveness of the Global Nuclear Detection Architecture (GNDA), specifically in the context of the following two tasks:

Task 1: Assess the feasibility of using performance measures and quantitative metrics for evaluating progress toward meeting the performance goals in the GNDA Strategic Plan.

The committee should assess the feasibility of using performance measures and quantitative metrics for evaluating progress in meeting these performance goals. This assessment should consider the following factors:

- Definition of performance measures for each of the performance goals in the Strategic Plan.
- Definition of quantifiable performance metrics for each performance measure including, as appropriate, efficiency, output, and outcome-oriented performance measures.
- Identification of data to be used to quantify these performance metrics.
- Identification of methodologies to be used to collect and analyze these data.
- Specification of performance target values for assessing the effectiveness of each performance measure.

If the use of performance measures and quantitative metrics is determined to be feasible, the committee should, to the extent practical, recommend specific performance measures, metrics, and the other supporting information described in the list above for consideration by the DNDO.

If the use of performance measures and metrics is determined to be infeasible, the committee should recommend alternative evaluation approaches.

Task 2: Recommend approaches for evaluating the overall effectiveness of the GNDA.

The committee should specifically recommend

- Approaches for developing an overall analysis framework to assess the effectiveness of the GNDA in terms of its ability to detect, deny, confuse, and/or deter adversaries.
- Approaches for exercising this analysis framework using combinations of modeling/simulation, red teaming, and/or related methods to assess the cost-effectiveness of and tradeoffs in GNDA components.

In executing these tasks the committee should examine efforts by other organizations to develop risk-informed metrics and analysis approaches for complex technological systems.

Appendix C

Biographical Sketches of the Committee

Arden L. Bement Jr., *Chair*, is the founding director of the Global Policy Research Institute and is the David A. Ross Distinguished Professor of Nuclear Engineering at Purdue University. He served as director of the National Science Foundation (NSF) from 2004 until 2010. Prior to his confirmation as NSF director, Dr. Bement served as director of the National Institute of Standards and Technology (NIST). He has held previous appointments at Purdue University in the schools of Nuclear Engineering, Materials Engineering, and Electrical and Computer Engineering, as well as a courtesy appointment in the Krannert School of Management. Dr. Bement joined the Purdue faculty in 1992 after a 39-year career in industry, government, and academia where his positions included vice president of technical resources and of science and technology for TRW, Inc.; deputy under secretary of defense for research and engineering; director, Office of Materials Science at DARPA; and professor of nuclear materials at the Massachusetts Institute of Technology. He served as a member of the U.S. National Commission for UNESCO and as the vice chair of the Commission's Natural Sciences and Engineering Committee. He is a member of the National Academy of Engineering, a fellow of the American Academy of Arts and Sciences, and a fellow of the American Association for the Advancement of Science. Dr. Bement has served on and has chaired numerous NRC studies, most relevant to this study chairing the Committee on the Scientific and Technical Assessment of Stockpile Stewardship. Dr. Bement is a retired lieutenant colonel of the U.S. Army Corps of Engineers and a recipient of the Distinguished Civilian Service Medal of the Department of Defense. He received his B.S. in engineering of metallurgy from the Colorado School of

Mines, his M.S. in metallurgical engineering from the University of Idaho, his Ph.D. in metallurgical engineering from the University of Michigan, and honorary doctorates from Cleveland State University, Case Western Reserve University, the Colorado School of Mines, the University of Idaho, the Korean Advanced Institute of Science and Technology, University of Macau, and the Michigan Technological University as well as a Chinese Academy of Sciences Graduate School Honorary Professorship. In 2013, Governor Mitch Daniels conferred the Sagamore of the Wabash Award from the State of Indiana to Dr. Bement. He has also been awarded the Order of the Rising Sun, Gold and Silver Star, from the Empire of Japan and the Chevalier dans l'Ordre National de la Légion d'Honneur from the French Republic.

Kelly Coyner is the executive director of the Northern Virginia Transportation Commission. As a senior researcher at the University of Maryland Center for Health and Homeland Security, Kelley Coyner formerly served as the chief of staff of the Senior Policy Group of the National Capital Region. In this capacity, she created and facilitated multiagency teams to oversee the selection, implementation, and evaluation of federally funded projects across all levels of government and more than 20 disciplines. She is an expert in the development and implementation of strategic planning for technically and socially complex fields; designing, implementing, and evaluating complex programs and policies across disciplines; and fostering innovation through public policy. Her previous work in emergency preparedness and response includes serving as the senior official in charge of emergency preparation and response at the Department of Transportation (DOT); civilian advisor to the U.S. Coast Guard Academy on coordinated disaster response (2001); visiting researcher at MIT's Center for Transportation Research (2001-2002); and observer, Executive Session on Domestic Preparedness at the Kennedy School of Government at Harvard University (2001-2003). Ms. Coyner's 6 years as a senior official at DOT included service as the Senate-confirmed administrator of the Research and Special Programs Administration. For her work in technology, transportation, and education, she received the Secretary's Gold Medal. From 2001 to 2006, Ms. Coyner lived in Bolivia and Paraguay, where she served as an advisor to local and international nonprofits on institutional development, conservation, community health, disaster relief, and risk mitigation. Ms. Coyner has served as an ex-officio member of the NRC's Transportation Research Board. She received her B.S. from Georgetown University's School of Foreign Service and the Dean's award for outstanding public service. A graduate of University of Virginia School of Law, Ms. Coyner clerked for Hon. George P. Kazen, Chief Judge (ret.), U.S. District Court, Southern District of Texas, and practiced with the Arent Fox law firm.

Martha Crenshaw is a senior fellow at the Center for International Security and Cooperation, a senior fellow at the Freeman Spogli Institute for International Studies, and a professor of political science by courtesy at Stanford University. She was the Colin and Nancy Campbell Professor of Global Issues and Democratic Thought and professor of government at Wesleyan University from 1974 to 2007. She chaired the American Political Science Association Task Force on Political Violence and Terrorism. She is a lead investigator at the National Consortium for the Study of Terrorism and Responses to Terrorism, a center of excellence of the Department of Homeland Security based at the University of Maryland. She has written extensively on the issue of political terrorism; in 2011, Routledge published *Explaining Terrorism*, a collection of her previously published work beginning with her 1972 article, "The Concept of Revolutionary Terrorism." She was a Guggenheim Fellow in 2005. Dr. Crenshaw has served on several National Research Council committees, most relevantly, the Committee on Determining Basic Research Needs to Interrupt the Improvised Explosive Device Delivery Chain. She received her B.A. from Newcomb College. She received her Ph.D. from the Department of Government Affairs at the University of Virginia.

James S. Dyer holds the Fondren Centennial Chair in Business in the College of Business Administration at the University of Texas, Austin. In 1999, he received the College of Business Administration Foundation Advisory Council Award for Outstanding Research Contributions. He served as chair of the Department of Information, Risk, and Operations Management for 9 years (1988-1997). He was the Philip J. Rust Visiting Professor of Business at the Darden Business School at the University of Virginia in 1999. He is the former president of the Decision Analysis Society of the Operations Research Society of America (now INFORMS). He received the Frank P. Ramsey Award for outstanding career achievements from the Decision Analysis Society of INFORMS in 2002. He was named a fellow of INFORMS in 2006 and also received the Multiple Criteria Decision Making Society's Edgeworth-Pareto Award in 2006. Dr. Dyer has consulted with a number of companies regarding the application of decision and risk analysis tools to a variety of practical problems, including the Jet Propulsion Laboratories, the RAND Corporation, and the Department of Energy. Dr. Dyer has published three books and more than 60 articles on risk analysis and investment science. His recent articles focus on decision making, including a multiattribute utility analysis for the disposition of weapons-grade plutonium in the United States and Russia. He received a B.A. with honors, Phi Beta Kappa, in physics with minors in mathematics and philosophy and his Ph.D. in business quantitative methods and management from the University of Texas, Austin.

Roger Hagengruber is director of the Office for Policy, Security and Technology, the Institute for Public Policy, and a professor of political science at the University of New Mexico. Dr. Hagengruber previously served as senior vice president at Sandia National Laboratories. During his 30 years at Sandia, he concentrated on nonproliferation issues and served as a negotiator in Geneva and the Soviet Union. Most recently, he coauthored the report: "Nuclear Power and Proliferation Resistance: Securing Benefits, Limiting Risk," which was published by the Nuclear Energy Study Group of the American Physical Society. His most recent and relevant National Research Council committee memberships were on the Committee on Evaluating Testing, Costs, and Benefits of Advanced Spectroscopic Portals and the Committee on Risk-Based Approaches for Securing the DOE Nuclear Weapons Complex. He received his B.S., M.S., and Ph.D. from the University of Wisconsin. His Ph.D. is in nuclear physics.

Capt. John M. Holmes, deputy executive director of operations at the Port of Los Angeles, oversees the Port Police, Port Pilots, Emergency Preparedness, Wharfinger, and Homeland Security divisions at one of the busiest container ports in the nation. He works cooperatively with associated government and law enforcement agencies to uphold maritime laws, enforce safety and security regulations, and continually test and enhance emergency response and preparedness procedures to ensure the safety of the Port workforce and residents in the surrounding harbor communities. Capt. Holmes has 30 years of international management experience in a variety of positions that include a chief operating officer, Fortune 500 executive, senior-level Coast Guard officer, and industry-renowned maritime security specialist. As Captain of the Port of Los Angeles, Holmes was at the helm on September 11, 2001, and has been credited with swift and decisive actions that ultimately led to the creation of a number of national security initiatives, including the Maritime Transportation Security Act (MTSA), Area Maritime Security Committee, and Sea Marshal Program. He most recently served as a principal and chief operating officer of the Marsec Group, a full-service security consulting firm specializing in supply-chain security, technology, and operations. Prior to forming the Marsec Group, Holmes was vice president and director of business development for Science Applications International Corporation, where he assisted government and commercial customers with the development of technological solutions to homeland security challenges, with an emphasis on port, border, and military solutions. His most relevant National Research Council committee membership was on the Committee on Evaluating Testing, Costs, and Benefits of Advanced Spectroscopic Portals. Capt. Holmes holds a B.A. in English and education from Boston College and an M.S. in business administration from Washington University's John M. Olin School of Business.

Edward H. Kaplan currently serves as the William N. and Marie A. Beach Professor of Management Sciences at the Yale School of Management, Professor of Public Health at the Yale School of Medicine, and Professor of Engineering in the Yale School of Engineering and Applied Sciences. He is an expert in operations research, mathematical modeling, and statistics and studies problems in public policy and management. His recent research has focused on counterterrorism topics such as the tactical prevention of suicide bombings, bioterror preparedness, and response logistics in the event of a smallpox or anthrax attack. He has received three Koopman Prizes from the Institute for Operations Research and the Management Sciences (INFORMS) Military Applications Society, one for his work on smallpox, another for his models evaluating suicide-bomber–detector schemes and most recently for his paper "Terror Queues," which was awarded in 2011. Dr. Kaplan codirects the Daniel Rose Technion–Yale Initiative in Homeland Security and Counterterror Operations Research. He served twice as the Lady Davis Visiting Professor at the Hebrew University of Jerusalem—in the School of Public Health and Community Medicine in 1994 and in the Department of Statistics in 1997—and is also an elected member of the Board of Governors of the Technion-Israel Institute of Technology. For all of his contributions to the operations research profession, Dr. Kaplan was designated an INFORMS fellow in November 2005 and is a member of both the National Academy of Engineering and the Institute of Medicine. He has served on numerous NRC committees, most relevantly on the Committee on Behavioral and Social Research to Improve Intelligence Analysis for National Security. He obtained his B.A. from McGill University with First Class Honors in economic and urban geography, and proceeded to graduate study at the Massachusetts Institute of Technology where he completed three masters' degrees (in operations research, city planning, and mathematics) in addition to his doctorate in urban studies.

John Mattingly recently joined the North Carolina State University (NCSU) where he is conducting basic research in radiation measurement and analysis methods applied to nuclear security applications, including nuclear materials control and accountability (NMC&A), arms control, safeguards, nonproliferation, counterterrorism, emergency response, and forensics. Prior to joining the NCSU faculty, he worked for 15 years at two national laboratories, Oak Ridge National Laboratories (ORNL) and Sandia National Laboratories (SNL). At ORNL, he helped to develop measurement systems and analysis methods for active neutron interrogation of special nuclear material for NMC&A, safeguards, arms control, and nonproliferation applications. This work included conducting experiments at several facilities in the U.S. nuclear weapons complex, and at foreign facilities including the Russian Institute of Experimental Physics (VNIIEF). At SNL, Dr. Mattingly was one of two lead developers of the Gamma Detector Response Analysis Software (GADRAS) which is commonly used by the on-call analyst community to simultaneously analyze gamma spectroscopy and neutron multiplicity counting measurements. At SNL, he also served as a 24/7 on-call analyst for the Department of Energy (DOE) and Department of Homeland Security. In this job, he learned to rapidly analyze gamma spectroscopy and neutron multiplicity measurements to assess the potential threat posed by a radiation source discovered in the stream of commerce and other settings. Dr. Mattingly received his B.S., M.S., and Ph.D. in nuclear engineering at the University of Tennessee.

Gregory S. Parnell is a visiting professor of industrial engineering in the Department of Industrial Engineering at the University of Arkansas. His research focuses on decision analysis, risk analysis, systems engineering, and resource allocation for defense; intelligence; homeland security; and environmental applications. He is also a senior principal with Innovative Decisions, Inc., a decision and risk analysis firm and has served as chairman of the board. Previously, he served as a professor of systems engineering at the U.S. Military Academy at West Point, a distinguished visiting professor at the U.S. Air Force Academy, an associate professor at Virginia Commonwealth University, and a department head at the Air Force Institute of Technology. Dr. Parnell is a former president of the Decision Analysis Society of the Institute for Operations Research and Management Science (INFORMS) and of the Military Operations Research Society (MORS). He has also served as editor of *Journal of Military Operations Research*. Dr. Parnell has published more than 100 papers and book chapters and was lead editor of *Decision Making for Systems Engineering and Management, Wiley Series in Systems Engineering* (2nd Ed, Wiley and Sons, 2011) and lead author of the *Handbook of Decision Analysis, Wiley Operations Research/ Management Science Series* (Wiley and Sons, 2013). He has received several professional awards, including the MORS Wanner Award, U.S. Army Dr. Wilbur B. Payne Memorial Award for Excellence in Analysis, MORS Clayton Thomas Laureate, two INFORMS Koopman Prizes, and the MORS Rist Prize. He chaired the NRC Committee on Methodological Improvements to the Department of Homeland Security's Biological Agent Risk Analysis (2008) and was a member of the Improving Metrics for the Department of Defense Cooperative Threat Reduction Program (2011). He is a fellow of MORS, INFORMS, the International Committee for Systems Engineering, and the Society for Decision Professionals. He received his B.A. in aerospace engineering from the State University of New York at Buffalo, his M.E. in industrial and systems engineering from the University of Florida, his M.S. in systems management from the University of Southern California, and his Ph.D. in engineering–economic systems from Stanford

University. Dr. Parnell is a retired Air Force colonel and a graduate of the Industrial College of the Armed Forces.

Donald Prosnitz is presently a senior principal researcher (adjunct) at RAND Corporation, a visiting scholar at the physics department of the University of California, Berkeley, and an independent technical consultant. His current activities include research on free-electron lasers and a range of studies at RAND concentrating on the utilization of technology to solve national and homeland security issues. Dr. Prosnitz was previously the deputy associate director (Programs) for Non-Proliferation, Homeland and International Security at Lawrence Livermore National Laboratory (LLNL) where he was responsible for overseeing all of the directorate's technical programs. He spent 2 years as an assistant professor at Yale University before joining LLNL. Over the next three decades, he conducted research on lasers, particle accelerators, high-power microwaves, free-electron lasers, and remote sensing, and managed the design, construction, and operation of numerous research facilities. In 1990, he was awarded the U.S. Particle Accelerator Award for Achievement in Accelerator Physics and Technology. In 1999, Dr. Prosnitz was named the first chief science and technology advisor for the Department of Justice (DOJ) by Attorney General Janet Reno. He was responsible for coordinating technology policy and technology projects among the DOJ's component agencies and with state and local law enforcement entities. In 2002, he was named a fellow of the American Physical Society. He is a former chair of the American Physical Society Forum on Physics and Society. He recently served on the NRC Committee to Review the Department of Homeland Security's Approach to Risk Analysis. Dr. Prosnitz received his B.S. from Yale University and his Ph.D. in physics from the Massachusetts Institute of Technology. He is a licensed amateur radio operator and an active member of his community's CERT (Community Emergency Response Team).

Thomas Schelling is a distinguished professor of economics at the University of Maryland. In 2005, he was awarded the Nobel Prize in Economics for enhancing the "understanding of conflict and cooperation through game-theory analysis." In 1990, he left the John F. Kennedy School of Government, where he was the Lucius N. Littauer Professor of Political Economy. He has also served in the Economic Cooperation Administration in Europe and has held positions in the White House and Executive Office of the President, Yale University, the RAND Corporation, and the Department of Economics and Center for International Affairs at Harvard University. Most recently, he has published on military strategy and arms control, energy and environmental policy, climate change, nuclear proliferation, and terrorism. Dr. Schelling is best known for his books *The Strategy* *of Conflict* and *Micromotives and Macrobehavior*. He is a member of the National Academy of Sciences, the Institute of Medicine, and a fellow of the American Academy of Arts and Sciences. In 1991, he was president of the American Economic Association, of which he is now a distinguished fellow. He was recipient of the Frank E. Seidman Distinguished Award in Political Economy and the National Academy of Sciences award for Behavioral Research Relevant to the Prevention of Nuclear War. He has served on numerous NRC committees, most relevantly the Roundtable on Social and Behavioral Sciences and Terrorism. Dr. Schelling received his B.A. in economics from the University of California, Berkeley, and his Ph.D. in economics from Harvard.

Detlof von Winterfeldt served as the director of the International Institute for Applied Systems Analysis (IIASA) in Laxenburg, Austria, prior to returning to the University of Southern California (USC), where he is a professor of industrial and systems engineering and a professor of public policy and management. Concurrently with his term at IIASA, he was a centennial professor in the Operational Research Group of the School of Management in the London School of Economics and Political Science. In 2003, he cofounded the National Center for Risk and Economic Analysis of Terrorism Events (CREATE) at USC, the first university-based center of excellence funded by the U.S. Department of Homeland Security. He served as CREATE's director until 2008. For the past 30 years he has been active in teaching, research, university administration, and consulting. His research interests are in the foundation and practice of decision and risk analysis as applied to technology development, environmental risks, natural hazards and terrorism. He is the coauthor of two books, two edited volumes, and author or coauthor of over 100 journal articles and book chapters on these topics. He is a former president of the Decision Analysis Society of the Institute for Operations Research and Management Science (INFORMS) and is an elected fellow of INFORMS and of the Society for Risk Analysis. In 2000, he received the Ramsey Medal for distinguished contributions to decision analysis from the Decision Analysis Society of INFORMS. In 2009, he received the Gold Medal from the International Society for Multicriteria Decision Making for advancing the field. He has served on numerous committees and panels of the NRC and the National Science Foundation; most relevant to this study was his service on the NRC's Committee on Transportation of Radioactive Waste. He received his B.S. in psychology with a minor in philosophy and his M.S. in psychology from the University of Hamburg, Germany, and his Ph.D. in mathematical psychology from the University of Michigan.

Appendix D

Model-Based Approaches for the GNDA

D.1 INTRODUCTION

Modeling and simulation are useful when

- the system/architecture performance is very important;
- the system/architecture is too complex to intuitively assess performance; and
- significant time and resources are required to improve performance.

The GNDA meets all of these conditions. Mathematical models have been used in many domains to support system/architecture design, performance evaluation, and resource allocation decision making. The purpose of this chapter is to present the potential for GNDA mathematical models to describe the architecture, evaluate the effectiveness, and support resource allocation decision making to increase GNDA effectiveness.

D.2 RATIONALE FOR GNDA MODELING

Mathematical models are developed for many different reasons. For example, sometimes models are derived as compact and precise statements of basic truths (e.g., physics). Sometimes models are created to explore the logical consequences of alternative conjectures about how certain systems behave (e.g., population biology). Sometimes models are employed to summarize statistical information about the past to create forecasts of the future (e.g., macroeconomics). And sometimes models are constructed to provide a framework for better decision making (e.g., operations research).

In thinking about how to evaluate the effectiveness of something as complicated as the GNDA, modeling has much to offer. First, by combining the detection characteristics of GNDA resources (e.g., sensors, human agents) with the physical deployment of such resources, it is possible to model the probability (and risk consequences) of nuclear material out of regulatory control entering the United States. Constructing such models serves the purpose of linking GNDA resource inputs and program activities to the primary outcomes of interest—interdiction and the risk consequences of failing to detect radiological or nuclear material. Second, such models can help identify appropriate performance measures for evaluating the effectiveness of the GNDA by identifying (via model analysis) the key variables that are associated with maximal detection and minimal risk. Third, models can help evaluate alternative hypotheses regarding effective GNDA design by comparing the modeled detection and risk outcomes of competing resource deployments. And fourth, by attaching appropriate costs to the different resources deployed, models can help identify the most efficient resource deployment at various budget levels.

The approach to modeling suggested above, with its emphasis on linkages between deployment of available resources and principal system objectives, has been and continues to be employed in support of major business and military decisions. By way of example, the next section reviews some applications of decision-oriented operations research modeling to selected military problems with the hope of convincing the reader that similar models could be developed to help evaluate the effectiveness of the GNDA.

D.3 EXAMPLES FROM ELSEWHERE

Modeling and simulation play an important role in defense analysis and support to decision makers. Military planning models exist at several levels: component, system, force structure (architectures), and campaigns. The models are used for force mission planning and force structure planning.

For example, consider strategic airlift to support a military campaign. Component models exist of aircraft airframes to estimate drag and fuel consumption. System models exist to calculate the time to deliver a plane's cargo to a destination. Force structure models exist to calculate the time to deploy a force (people and equipment) for a mission. Finally, campaign models exist to determine the time to achieve the campaign objectives given the force available and the potential actions of the adversary. For force structure planning, the models are used to help determine the best mix of aircraft (e.g., tactical and strategic) and an affordable amount of airlift capability given the potential threats our nation might face on the strategic planning horizon. In both cases, the models do not make decisions but, rather, they inform the analysts, strategic planners, and decision makers. Additional examples can be found within the report.

D.4 GNDA-SPECIFIC CONCERNS

Any focused modeling application must be responsive to the specifics of the system under study. This is certainly true of the GNDA. As discussed elsewhere in this report, the GNDA is a three-layered architecture—the internal (or domestic) layer, the U.S. border layer, and the international layer. Responsibility for each layer rests with different agencies. It is therefore convenient to think about detection/interdiction and risk consequences as a function of GNDA activities and resource deployments within each of these three layers first, and then use the layer-specific analyses to create an overall model for the GNDA.

The GNDA does not have a centrally managed budget. Rather, each of the agencies that participates in the GNDA determines which of its activities qualify as GNDA-related, what resources are devoted to those activities, and how much they cost. These determinations are historical in nature, that is, "after-the-fact" estimates of how much money was spent on various GNDA activities. Although these agency contributions have been totaled to produce what looks like the total GNDA budget, there is no prospective procedure that determines how much money the government allocates to the GNDA. The GNDA thus operates under what could be called an best-effort budget. This is important to note, because in thinking about the GNDA, it may not be possible to optimally allocate the resources of this best-effort budget across different activities in a way that would change the contributions of participating agencies to the overall budget. This does not imply that modeling optimal GNDA resource allocation is without purpose, however, because the gap between extant and optimal resource allocation will provide a measure of the *cost* of operating the GNDA in the manner chosen.

Another GNDA-specific concern is that the risks the system is trying to minimize (nuclear materials out of regulatory control entering the United States) derive in the main from the actions of intelligent adversaries such as terrorists or hostile states. If such adversaries are intent upon attacking the United States with nuclear or radiological materials, then surely they will adapt their behavior given changes in GNDA resource deployments to better achieve their goals. There are different choices possible for how such adaptive behavior should be modeled. Traditional game theory models adopt a "worst case" viewpoint that essentially grants adversaries perfect foresight, while less pessimistic approaches presume that potential attackers know some things but not others about GNDA activities (or know about resource deployments of different assets with different probabilities). None of these frameworks are comparable to the "human vs. nature" models that characterize risk analysis for naturally occurring threats such as floods, earthquakes, or epidemics, for example, and so it is important to think hard about the adversarial nature of the risks that the GNDA seeks to mitigate.

D.5 POTENTIAL MODELING APPLICATIONS FOR THE GNDA

In the section that follows the committee develops some simple examples for the purpose of illustrating insights that one can gain from modeling and, at a basic level, some of the thought processes involved. It is not meant to provide a complete set of models that should be developed. Later, we discuss the availability of and need for more advanced modeling methods to help evaluate the GNDA.

D.5.1 Descriptive Modeling

The first task when modeling any system is to understand the basic relationships among inputs, processes, and outputs. Such models are descriptive in nature, are meant to help understand how the system in question actually works, and also serve as building blocks for downstream decision-oriented models that address resource allocation or other issues.

In thinking about the GNDA, one set of descriptive models would seek to answer the following basic question: Given a particular physical deployment of agents and sensors in a particular setting (e.g., a port, border crossing, along a highway), what is the likelihood that the entry of nuclear or radiological material out of regulatory control into the area of interest would be detected? As an extremely simple example, suppose that each of n sensors is capable of detecting a threat with probability p, and that detection is independent across sensors. Then the probability that a threat would be detected via the deployment of n sensors would equal $1 - (1 - p)^n$, a graph of which appears in Figure D-1 the assumption that each sensor detects with probability 0.2.

D.5.2 Sensor Quality

One of the Domestic Nuclear Detection Office's (DNDO's) missions is to develop new sensor technologies (see DNDO acting director's statement to Congress in July 2012).[1] Once we have a descriptive model, we use the model as a tool to evaluate potential new sensor capabilities by assessing the impact on the system performance measure, probability of detection, of

[1] The written statement from Dr. H.A. Gowadia, DHS, provided July 26, 2012 can be accessed at http://homeland.house.gov/sites/homeland.house.gov/files/Testimony-Gowadia.pdf.

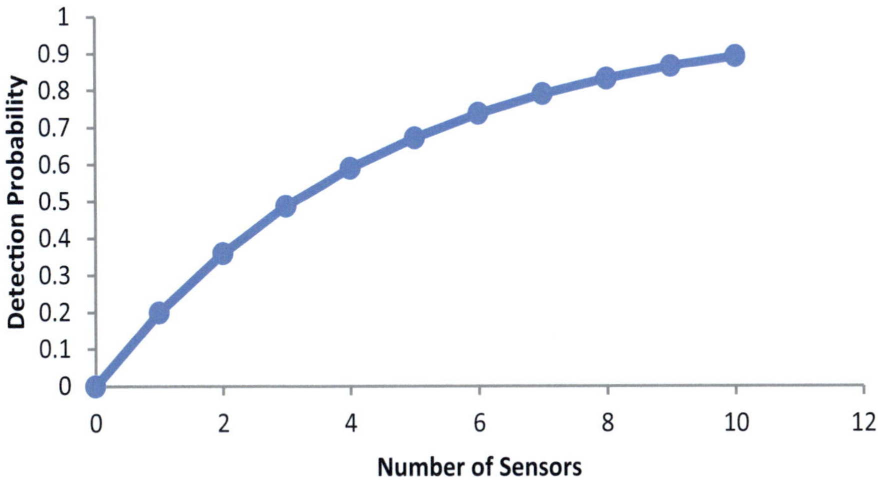

FIGURE D-1 Detection Probability as a function of the number of deployed sensors. This is an illustration of diminishing returns, in that doubling the number of sensors increases the detection probability by less than a factor of 2. The model is perhaps the simplest that can be envisioned, but the important point is to see how a model links inputs (the number of sensors) to outputs (in this case the probability of detection).

the extant systems with potential sensor improvements. Returning to our sensor system model, above, with a probability of detection of 0.2, suppose we want to assess the of system capability improvement of increasing the probability of detection to 0.3 or 0.4. Suppose the system goal was a probability of detection of 0.8. How many of each sensor would be required to achieve the goal?

To achieve a system probability of detection of 0.8, we would require eight of the $P = 0.2$ sensors, five of the $P = 0.3$ sensors, and about three of the $P = 0.4$ sensors. The model in Figure D-2 is very simple; it is intended to show how a model links inputs (the number of sensors and sensor performance) to outputs (in this case the probability of detection).

D.5.3 False Positive Versus False Negative Errors

When we model sensors we need to consider false negative and false positive errors. For GNDA, a false negative error is the probability of not detecting nuclear or radioactive material when it is present; that is, the sensor does not alarm when threat material is present (also called a missed detection or a false negative). A false positive error is the probability of a

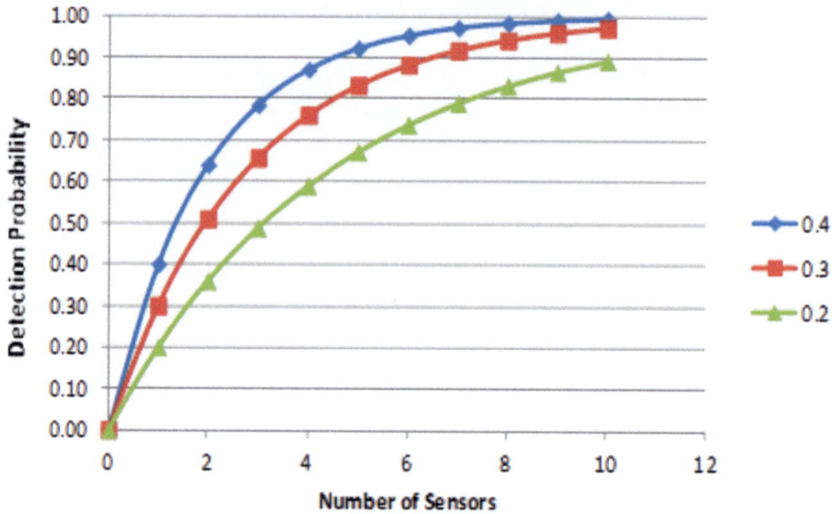

FIGURE D-2 Detection Probability as a function of the number of deployed sensors of varying performance. Like Figure D-1, this figure illustrates diminishing returns, but more rapidly for the better-performing sensors.

false detection; that is, the sensor indicates detection when the material is not present in sufficient quantity. While GNDA is primarily concerned with minimizing false negative errors for preventing the illicit transport of nuclear or radiological material, false positive errors can significantly increase the detection cost and impose a burden on the organization whose vehicle or container created the a false detection. Unfortunately, false positive and false negative errors cannot be simultaneously reduced. For example, if we lower the detection threshold to reduce the probability of missing a valid detection (a false negative error), we increase the probability of false detection (false positive errors).

The number of false positive errors can be significant for low-prevalence events. A numerical example can help to illustrate the magnitude. Let T and NT be the presence or absence of, respectively, the nuclear/radioactive materials that could be a threat. Let D and ND refer to the probability that the sensor detects (alarms) or does not detect the material (does not alarm).

Suppose we are provided the data in Table D-1. The detection probabilities seem to be quite high. The probability of a false negative error is 0.01 and the probability of a false positive error is 0.05. Suppose that 1 in

TABLE D-1 Numerical Example Illustrating the Magnitude of False Negative and False Positive Errors for a Low-Prevalence Event

		False Negative		False Positive	
	P(D I T]	*P*(ND I T]	*P*(ND I NT]	*P*(D I NT]	*P*[T]
Sensor	0.99	0.01	0.95	0.05	

where
D = detect
ND = not detect (Type I)
T = threat present
NT = no threat present
P(D I T) = probability of detecting a potential threat
P(ND I T) = probability of not detecting a potential threat (a false negative error)
P(ND INT) = probability of not detecting a non-threat (not alarming on a non-threat)
P(D INT) = probability of detecting a non-threat (a false alarm error)

100,000 inspections contains the threat material.[2] Given we have a detection, what is our probability that we found the threat material?

This calculation can be done with Bayes' Law; however, we will use a simpler calculation. Suppose there are 10,000,000 inspections. On average, there would be 100 inspections that find the threat material. The sensor would properly detect 99 of these and miss 1. However, of the 9,999,900 inspections without the threat material, 5 percent of the time or 499,995 detections would be false positives. Therefore, for any given detection, the probability of having the threat material would be only 0.02 percent [99/(99 + 499,995)].

D.5.4 Resource Allocation Given Extant GNDA Capabilities

As discussed earlier, the GNDA functions with a best-effort budget that precludes efficient resource allocation and substitution; yet, within agencies or jurisdictions of different agencies participating in the GNDA, some flexibility is possible. Continuing with our simple example, suppose that a geographic area is subdivided into two zones A and B, and the participating GNDA agency is trying to decide how many of its 10 sensors it should deploy in zone A versus zone B. From current intelligence assessments, the agency believes that if an adversary were to attempt to bring illicit nuclear or radiological material into the area of interest, there is a conditional probability a that entry would occur in zone A and a complementary conditional probability $b = 1 - a$ that entry would occur in zone B. This being the case, if the agency deployed n sensors in zone A and $10 - n$ sensors in zone B,

[2] In an actual modeling study, this value would be determined by intelligence estimates.

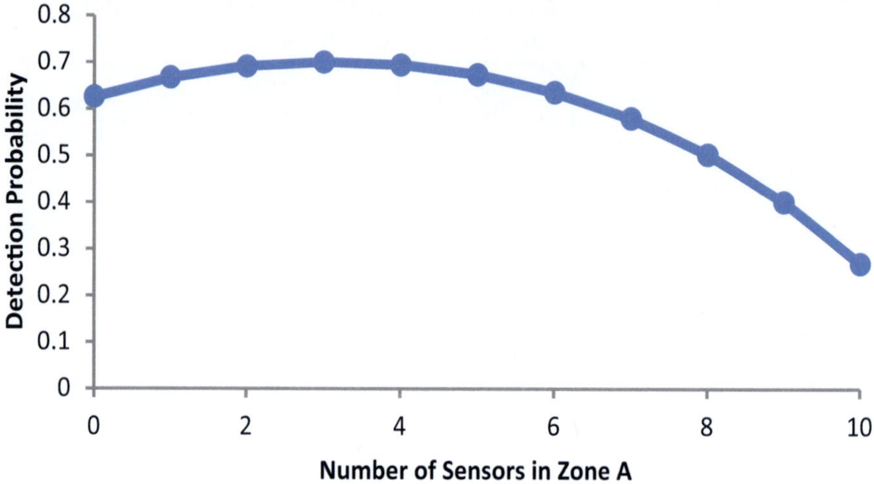

FIGURE D-3 Detection probability as a function of the number of sensors allocated to zone A (n) for the case where a = 0.3 (and b = 0.7) and P = 0.2.

then detection upon entry would occur with probability $a[1 - (1 - P)^n] + b[1 - (1 - P)^{10-n}]$. A graph of this detection probability as a function of the number of sensors allocated to zone A (n) appears in Figure D-3 for the case where a = 0.3 (and b = 0.7) and, as before, P = 0.2. The key feature of this graph is that the probability of detection is highest when three sensors are placed in zone A and seven in zone B, which results in an overall detection probability of 70 percent. Clearly there are many ways of deploying the 10 sensors. This example shows how it is possible that even under the best-effort budget of extant resources it is possible to think about different deployments to improve the likelihood of detection or cost-effectiveness.

Cost-effectiveness is usually defined as the incremental (additional) cost required to achieve an incremental unit of performance. Operationalizing cost-effectiveness requires a measure (or measures) of effectiveness, and a costing model for the level of performance (effectiveness) that can be reached at different resource levels. To illustrate, consider the previous example that shows how to best allocate 10 sensors between two zones. Suppose that each sensor cost $100k/year to operate, so in total, $1M/year is being spent, and suppose also that there is one infiltration attempt per year. To maximize cost-effectiveness in this case means to minimize the cost per detected infiltration, which of course is achieved by maximizing the probability of detection. The example shows that placing three sensors in zone A and seven in zone B maximizes detection probability at 70 percent, thus, the cost per detected event equals $1M/.7 = $1.43 million per case detected. Any other allocation would have a lower detection probability and thus a higher cost per detected event. For example, putting eight sensors in zone A and two in zone B would yield a detection probability of 0.5 and thus a cost per detected event of \$1M/.5 = \$2M. Clearly the first allocation is more *cost-effective* than the second.

Estimating the cost-effectiveness of the entire GNDA is a challenging task, because it requires determining (or more likely modeling) the cost and the effectiveness of alternative allocations of GNDA resources. Nonetheless, the principle is the same as in the simple example above.

D.5.5 Sensitivity Analysis

In many modeling applications, we may not be certain of the expert data, especially if the experts do not have a large number of historical incidents to assess. Suppose in our previous example, the intelligence analyst believed the probability of attack in zone A (*a*) could be 0.3 to 0.5. We can easily use our model to assess how sensitive our model is to that input data.

Figure D-4 shows the sensitivity of the probability of detection to the intelligence analysis assumption about the probability of attack in zone A versus zone B.

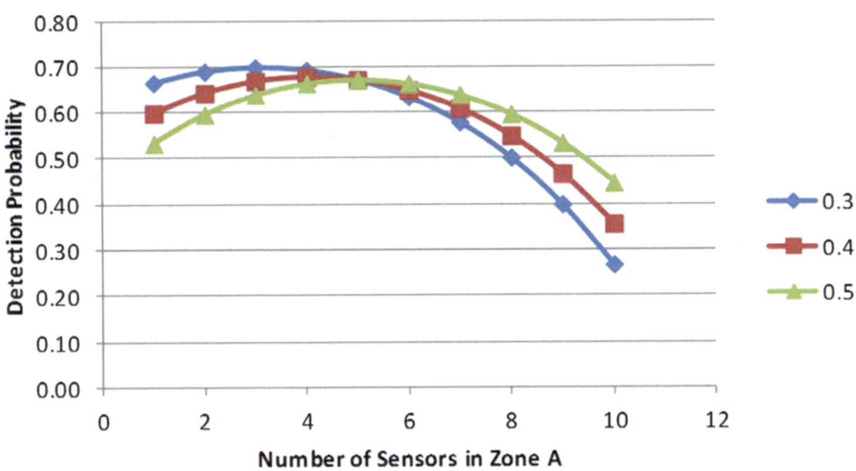

FIGURE D-4 Detection probability in two-zone, 10-sensor example illustrating the sensitivity of detection probability to intelligence estimates of probability of attack within a particular zone.

D.5.6 Intelligent Adversary

The example above treated the behavior of a terrorist as not reacting to GNDA actions to deploy sensors within zones A or B the same way one thinks of a weather forecast—there is a 30 percent chance of rain, and by the way, there is a 30 percent chance that illicit nuclear materials will be smuggled into zone A. Perhaps a more realistic approach is to recognize that if terrorists (or operatives from a rogue state) are intent upon bringing nuclear material into the country for the purpose of mounting an attack, such operatives are likely to have studied our defensive posture so that they can commit to a plan that is, from their vantage point, most likely to succeed. To see how such models can be constructed, suppose that the sensors in question are overt and easily observed (as is the case in large ports, for example). Then, from the defenders' point of view, the worst case is that the terrorists know how many sensors are allocated to zone A versus zone B. In this game, the terrorists seek to minimize the chance that they will be detected. Thus, given any split of the sensors between zones A and B and assuming the terrorists are aware of the split, the terrorists will *choose* the zone with the lowest probability of detection, and the chance that the defender would detect entry reduces to the chance of detection in the zone with the fewest sensors.

If four or fewer sensors are deployed in zone A, then the terrorists will select zone A, but if six or more sensors are deployed in zone A, the terrorists will select zone B. The result that is best for the government, and at the same time worst for the terrorists, is to place five sensors in each zone. This serves to equalize the likelihood that the illicit materials will be detected; it is 67 percent in either zone (see Figure D-5) . With this result, it does not matter if the terrorists select to infiltrate zone A or zone B (or if they choose to flip a coin to choose between A and B). This turns out to be a much more general proposition—when defending against intelligent adversaries, worst-case analysis requires defenders to *minimize* the terrorists' *maximum* probability of success (or more generally the expected risk consequences of terrorist success including, for example, morbidity, mortality, economic, and political damage). To achieve this, defenders must *equalize* the payoffs to the terrorists across their various options. Achieving such equalization in payoffs produces a certain robustness, in that the likelihood and consequences of terrorist success are fixed no matter what the terrorists decide to do.

Again, to understand why this result must be correct, note that if the different terrorist options produce different payoffs, then just as in our example with zone A and zone B, terrorists will gravitate toward their most attractive option, which will be more rewarding to them (and damaging to us) than what can be achieved from equalizing the payoffs. Worst-case defense also provides a certain level of comfort when thinking about terror-

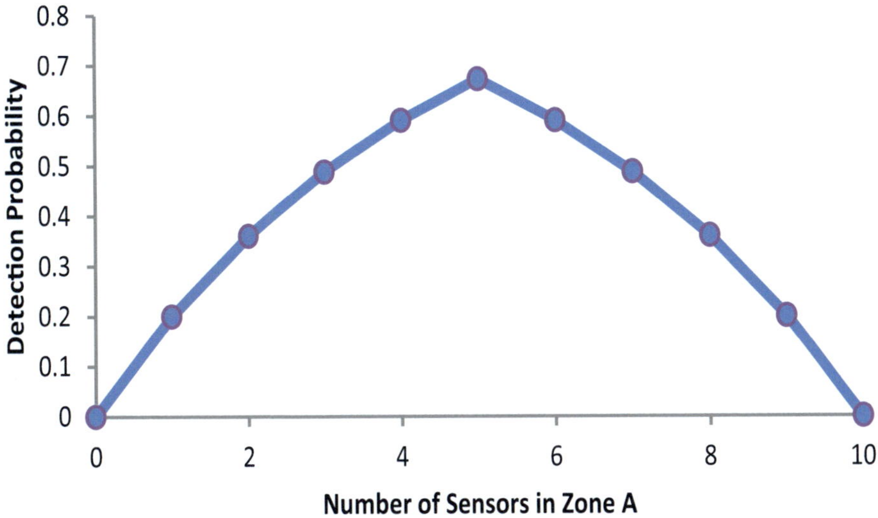

FIGURE D-5 Detection probability for the two-zone, 10-sensor example when an adversary has knowledge of resource allocation.

ism, because if it turns out that the terrorists are not as smart as imagined (i.e., they cannot see our defenses perfectly and hence cannot choose optimally themselves), then whatever results actually occur will be less severe than conjectured in the modeling analysis.

However, for defenders to take advantage of terrorists' lack of information requires "knowing just what they don't know" (see Section 4.2 within the main body of the report). In the two-zone example above, if the defenders strongly believed that terrorists were likely to attack zone A with probability 0.3 and hence placed three sensors in A and seven in B, the defenders would think (from the analysis in the second example) that they would detect with probability 70 percent. But if the defenders were wrong in their assessment, smart terrorists would choose to infiltrate zone A, which contains only three sensors, with certainty (e.g., using insider information), so the probability of detection would shrink below 50 percent.

D.5.7 Extension to Risk Consequences and Randomization Defense

To illustrate how the ideas above extend beyond the probability of detection to risk consequences, consider Table D-2. For simplicity we presume that there are only four defensive agents (which could be human agents, sophisticated sensors, or both working together) to defend zones A and B. The table reports expected casualties in zone A or B as a function of the

TABLE D-2 Extension to Risk Consequences

# Units at A	Casualties(A)	Casualties(B)	Terrorist Attacks
0	32	6	A
1	16	7	A
2	8	9	B
3	4	12	B
4	2	16	B

number of units (or defensive agents) deployed to zone A, along with the choice made by an intelligent terrorist. The terrorist, who observes the allocation and concludes that the best the defender seems able to do is to allocate two agents to each zone, would induce an attack in zone B with an expected nine casualties.

This seems inconsistent with the previous example where we argued that the defender should seek to equalize the attackers' payoffs across the choices they face; in the present example, after allocating two agents to each zone, the attackers would expect eight casualties in A and nine in B, hence their decision to attack B. Is it possible for the defenders to do better? The answer is yes, and randomization provides the key. Suppose that instead of committing two agents to defend each of zone A and B, the defender randomized with probability 0.9 that two agents are assigned to zone A and two to zone B, and with probability 0.1 that one agent is assigned to A but three to B. As illustrated in the decision tree (Figure D-6), this randomization equalizes the expected casualties in zones A and B to 8.8, a modest reduction over the fixed deployment of two agents to each zone.

Should zone A be attacked, the expected casualties equal $0.9 \times 8 + 0.1 \times 16 = 8.8$, whereas if zone B is attacked, expected casualties are given by $0.9 \times 9 + 0.1 \times 7$, which again equals 8.8. Randomization is thus a powerful mechanism for defending against strategic attackers. We note that randomization is already employed in homeland defense: U.S. Air Marshals are randomly rotated across flights, while defensive patrols at Los Angeles Airport are also randomized to better defend against terrorist attacks (Jain et al., 2010).

D.5.8 Resource Allocation Modeling

In the examples above, the allocation decisions faced by defenders all involved the placement of different numbers of otherwise equivalent sensors or agents in different zones. Recall that such examples were motivated by

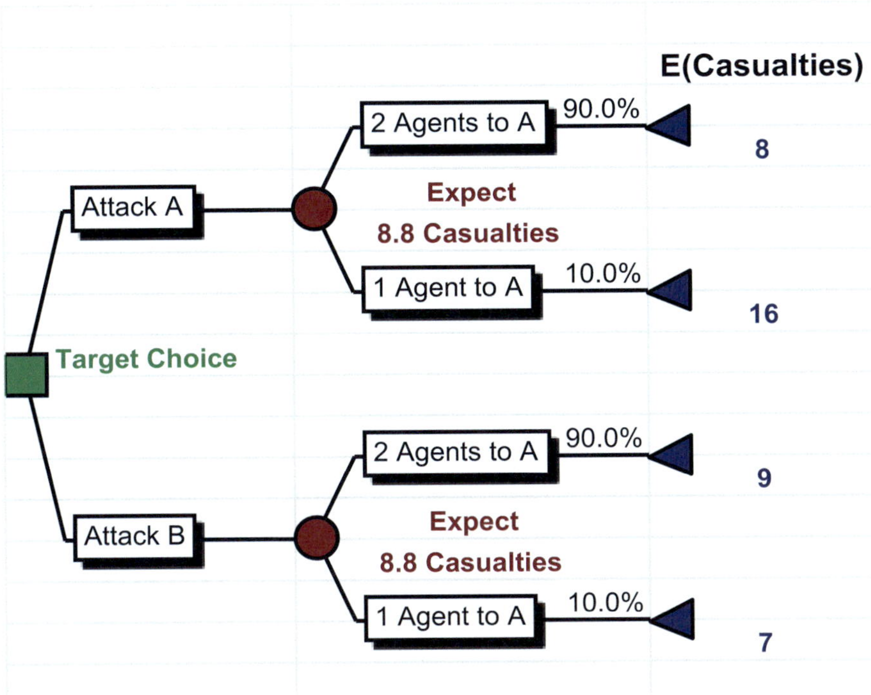

FIGURE D-6 Decision tree illustrating that randomization equalizes the expected casualties in zones A and B to 8.8.

the idea that within a given jurisdiction, some lead agency with fixed physical resources (such as agents or sensors) might still face some flexibility in how to deploy those resources. To see how the same resource allocation logic can apply when there are multiple resources with different unit costs available for GNDA use, suppose that there are some number m of different resource types (e.g., sensors with different costs and different sensitivity and specificity [equivalently different likelihoods of committing false negative and false positive errors as discussed previously]), and that there are also some number n of distinct detection or interdiction tasks (where "task" could mean "detect a specific threat type" or "detect any threat in a given geographical location"). Let x_{ij} denote the number of units of type i resources that are allocated to task j. We refer to x_{ij} as *decision variables* because assigning numbers to such variables is equivalent to deciding how many type i resources to allocate to task j. If c_{ij} represents the cost of allocating one unit of resource i to task j, then $c_{ij}x_{ij}$ is the cost of this particular decision. Summing $c_{ij}x_{ij}$ over all i (between 1 and m) and j (between 1 and

n) then yields the total cost of making all decisions that allocate resources to tasks. Presumably this sum cannot exceed the total budget available for the set of tasks under consideration. Also, because we are dealing with physical resources, each of the decision variables x_{ij} must be nonnegative, while there are likely additional constraints governing the total number of resources available by type. A proposed resource allocation plan as implied by assigned numerical values of the decision variables is said to be feasible if its total cost resides within the available budget and if no other constraints on resource availability are violated.

Now, in similar spirit to the simple models discussed earlier, the likelihood of interdicting an attempted infiltration with nuclear material out of regulatory control (or more generally the expected risk consequences of any terrorist infiltration plan) can be estimated using more complex models. Models that allow for the behavior of intelligent adversaries can also be developed. Again, the goal of such models is to produce a set of resource allocation decisions that are likely to lead to good (if not optimal) GNDA outcomes.

Now, recall our earlier discussion of GNDA's best-effort budget. Optimal resource allocation as suggested by models of the form above could result in radically different suggestions for how to best allocate the total amount of money spent on GNDA activities. Using these same models in descriptive mode—that is, setting the decision variables equal to the values implied by current GNDA operations—provides an immediate basis for comparison: for the same total amount of money spent, how much better would optimal resource allocation perform than current practice? Equivalently, what is the penalty paid for operating the GNDA in its current "pure participation" mode when compared with the protection offered from optimally disbursing the total GNDA budget? What loss in the likelihood of detection (or other risk consequences) results from forcing the GNDA to operate under a best-effort budget as opposed to rationally allocating a fixed central budget?

This logic can also be applied to the GNDA's three layers—each of the domestic, border, and international layers has an associated best-effort budget that equates to the total amounts spent by all participating agencies in layer-specific GNDA activities. How might the prospects for detection and risk reduction improve if each of these budgets was centrally allocated? And, thinking across these layers, how much further could system performance improve if the amounts allocated to each layer were allowed to vary (e.g., across detection or transportation modalities) so as to maximize safety from the threat of nuclear or radiological terrorism?

Appendix E

Acronyms

BMD	ballistic missile defense
CBA	capability based assessments
DHS	Department of Homeland Security
DNDO	Domestic Nuclear Detection Office
DNI	Director of National Intelligence
DOD	U.S. Department of Defense
DODAF	Department of Defense Architecture Framework
DOE	U.S. Department of Energy
DOJ	U.S. Department of Justice
DOS	U.S. Department of State
DTRA	Defense Threat Reduction Agency
EPA	Environmental Protection Agency
FAA	functional area analysis
FBI	Federal Bureau of Investigation
FNA	functional needs analysis
FY	fiscal year
GAO	Government Accountability Office
GNDA	Global Nuclear Detection Architecture
GPRA	Government Performance and Results Act

HEU	highly enriched uranium
IND	improvised nuclear device
NCT	nuclear counterterrorism
NLE	national-level exercises
NNSA	National Nuclear Security Administration
NRC	National Research Council
OMB	Office of Management and Budget
OSD	Operations Support Directorates
OUO	Official Use Only
OV	operational view
PRND	preventive radiation and nuclear detection
RDD	radiological dispersal device
RN	radiological and nuclear
RNTRA	radiological and nuclear threat risk assessment
SAFE Port	Security and Accountability for Every Port
SOP	standard operating procedure
V&V	validation and verification
WMD	weapon of mass destruction

Appendix F

Glossary

The terms defined in this glossary are for the purpose of usage within this report.

Adaptive adversary An opponent that continually reacts to defensive actions and protective measures by adapting or inventing new pathways to do harm.

Analytic framework An analytic framework is a consistent way of thinking in a planning or decision-making context, which usually involves specific analytical tools (e.g., cost-benefit analysis, risk analysis, Bayesian analysis) and methods (e.g., statistics, expert elicitation, value of information analysis).

Architecture Often used in connection with the term "systems" as in "systems architecture." A (systems) architecture consists of basic elements that interrelate and connect to form subsystems and the system as a whole. The definition of an architecture also includes the purpose of the system and its boundaries.

Best-effort budget An overall value that represents the GNDA budget; it is determined by reports from each of the agencies that participates in the GNDA. Each agency estimates the activities, resources and costs that are GNDA-related.

Capability The ability to perform tasks or assessments based on available capacities that lead to outcomes. Capacity (see below) and capability together describe the resources in place to achieve a predefined task. For example, a vaccination program may have a capacity of 10 million vaccines which would lead to the capability of vaccinating 1 million people per day.

Capacity The tools and core resources that enable action; for example, detectors, concepts of operation, and training are capacities.

Detection The action or process of determining the presence of a target object or substance, by way of passive and active detection equipment as well as by nontechnical means, such as ongoing law-enforcement information or observations, public and other government departments' and agencies' observations, or reporting of suspicious behavior (GNDA, 2010).

Deterrence Reduction by denial, by retribution, by other means of the likelihood that adversaries will attempt an attack.

Gap A pathway without detection capabilities (as used within the GNDA).

Goal A statement of the result or achievement toward which effort is directed. Strategic goals articulate clear statements of what the agency wants to achieve to advance its mission and address relevant national problems, needs, challenges, and opportunities. These outcome-oriented strategic goals and supporting activities should further the agency's mission (OMB, 2012).

Goals are usually defined as specific attainment level or target on an *objective* or *performance measure*. In some contexts (e.g., GNDA, 2012; NAS, DTRA) the term goal is used synonymously with a higher-level objective.

Implementation plan An implementation plan defines the specific programs and steps to be taken to implement a strategic plan. It is usually more short term (1-3 years) than a strategic plan and it defines specific activities and desired goals and targets.

Indicator A measurable value that is used to track progress toward a goal or target. (See "metric" below.) "Agencies are encouraged to use *outcome indicators* . . . where feasible" (OMB, 2012).

Metric Synonymous with "indicator," the actual quantity that is used to measure progress. Metrics can be quantitative or qualitative. Quantitative metrics may use numerical (e.g., a percentage or number) or constructed scales (e.g., high, medium, low). A standard or indicator of measurement (OMB, 2012).

Measure Qualitative or quantitative facts that gauge the progress toward achieving a goal. These facts may be in the form of indicators, statistics, or metrics.

Mission boundary The boundary and transfer of federal responsibilities across the spectrum of nuclear counterterrorism activities.

Mission statement A brief, easy-to-understand narrative . . . [that] defines the basic purpose of the agency and is consistent with the agency's core programs and activities expressed within the broad context of national problems, needs, or challenges" (OMB, 2012).

Objectives (or Strategic objectives) The outcome or impact the agency is trying to achieve (OMB, 2012).

Out of regulatory control Materials that are being imported, possessed, stored, transported, developed, or used without authorization by the appropriate regulatory authority, either inadvertently or deliberately (DHS, 2011b, Vol. I, p. 4).

Outcome A type of measure that indicates progress against achieving the intended result of a program. Indicates changes in conditions that the government is trying to influence (OMB, 2012).

Outcome-based Provides information on progress made against an intended result or can indicate changes in conditions that the customer is attempting to influence.

Output Description of the level of an activity or a program, such as the number of inspections conducted in a given day or the number of interagency meetings held per year. Outputs may not necessarily be related to outcomes.

A type of measure, specifically the tabulation, calculation, or recording of activity or effort, usually expressed quantitatively. Outputs describe the level of product or activity that will be provided over a period of time. While output indicators can be useful, there must be a reasonable connection between outputs used as performance indicators and outcomes (OMB, 2012).

Performance goal Goals established by an agency to monitor and understand progress toward strategic objectives (OMB 2012).

A statement of the level of performance to be accomplished within a time frame, expressed as a tangible, measurable objective or as a quantitative standard, value, or rate. For the purposes of this guidance and implementation of the GPRA Modernization Act, a performance goal includes a performance indicator, a target, and a time period. The GPRA Modernization Act requires performance goals to be expressed in an objective, quantifiable, and measurable form unless agencies in consultation with OMB determine that it is not feasible. In such cases an "alternative form" performance goal may be used. The requirement for OMB approval of an alternative-form goal applies to performance goals only. Milestones are often used as the basis of an alternative-form performance goal. Performance goals specified in alternative form must be described in a way that makes it possible to discern if progress is being made toward the goal (OMB 2012).

Performance measure "[M]easurable values that indicate the state or level" and that are "used to track progress toward a goal or target. . . . By definition, the indicators for which agencies set targets with time frames are performance indicators. . . . Agencies are encouraged to use *outcome indicators* as performance indicators where feasible" (OMB, 2012).

Performance metric A performance metric is a specific prescription of how to make an estimate or assessment of a decision alternative or plan on a *performance measure*. For example, the *performance measure* "Life-cycle cost" can be estimated by the metric "discounted life-cycle cost in 2012 dollars." Often the metric is defined as part of a *performance measure*.

Port(s) of Entry The terms "port" and "port of entry" incorporate the geographical area under the jurisdiction of a port director (19 CFR § 101.3).

Proxy A metric that does not directly relate to a goal or objective but can be used as an indirect measure as long as a strong relationship between the metric and its objective can be made. Proxies can be useful and should not be indiscriminately avoided especially when a direct metric cannot be established. Proxy metrics are also called "indirect metrics."

Scope (of a goal, objective, or metric) The focus or functional scope consists of three main levels as they relate to the GNDA:

Architecture—the integrated capability of all three geographic layers and the crosscutting functions of the GNDA.

Layers—the operational elements and assets in each of the three geographical layers of the GNDA (exterior to the United States, border, and interior of the United States).

Resources—budgets, people, assets, and capabilities.

Strategic goals The general, *outcome-oriented*, long-term goals for the major functions and operations of the agency (OMB, 2012).

Strategic objectives Objectives that reflect the *outcome or impact* the agency is trying to achieve (OMB, 2012).

Strategic plan Presents the long-term objectives an agency hopes to accomplish, set at the beginning of each new term of an Administration. It describes general and longer-term goals the agency aims to achieve, what actions the agency will take to realize those goals, and how the agency will deal with the challenges likely to be barriers to achieving the desired result. An agency's strategic plan should provide the context for decisions about performance goals, priorities, and budget planning, and should provide the framework for the detail provided in agency annual plans and reports (OMB, 2012).

Surge A domestic operation under DNDO authority to be a temporary increase in intensity of preventive detection activities using existing limited resources, initiated by weak intelligence cues that do not constitute the basis for determining a credible threat that would otherwise trigger search operations (JASON, 2012, p. 9).

Type There are different types of goals, objectives, and metrics.

Input—indicates consumption of resources used (e.g., time, money).

Process—indicates how well a procedure, process, or operation is working.

Output—describes the level of activity (or product) that will be provided over a period of time.

Outcome—indicates progress against achieving the intended result.

Vision statement Describes the long-term goal of the organization or program (OMB 2012).